47 Structure and Bonding

Ferrites
Transition Elements
Luminescene

With Contributions by
I. K. Fidelis A. P. Golovina J. C. Grenier
P. Hagenmuller T. J. Mioduski
M. Pouchard V. K. Runov N. B. Zorov

With 40 Figures and 12 Tables

Springer-Verlag
Berlin Heidelberg GmbH 1981

DOI 10.1007/978-3-540-38705-3

Originally published by Springer-Verlag Berlin Heidelberg New York in 1981
MyCopy version of the original edition 1981

2152/3140-543210
www.springer.com/mycopy

Table of Contents

Vacancy Ordering in Oxygen-Deficient Perovskite-Related Ferrites

Jean-Claude Grenier, Michel Pouchard and Paul Hagenmuller

Laboratoire de Chimie du Solide du CNRS, Université de Bordeaux I, 351, Cours de la Libération, F-33405 Talence Cedex, France

Structure and Bonding 47

1 Introduction

During the last thirty years, much work has been carried out in the field of the non-stoichiometry of transition-metal oxides. These studies were generally purely structural, using X-ray diffraction and, more recently, transmission electron microscopy has been applied. Many models for non-stoichiometry have been proposed [1,2]. Only recent developments in solid state chemistry have correlated the physical properties and the chemical features with the nature and grouping of the anionic defects.

For very small defect concentrations, an interpretation was given for the first time by Frenkel, Schottky and Wagner who considered the point defects as being randomly distributed in the lattice without interaction. This theory was based on a statistical thermodynamic treatment[3]. Later, Magnéli and Wadsley showed that at higher concentrations, the non-stoichiometry generally leads to an ordering of the defects, which tend to associate in order to minimize the lattice energy[4-6]. The network is thus rearranged. Depending on the material and the defect ratio, different kinds of distributions appear. Thus, clusters (for example in UO_{2+x}) or microdomains (as in the $Fe_{1-x}O$ type oxides) may be observed. If long-range ordering occurs, new compounds with definite compositions and structures can be obtained. This is the case with some non-stoichiometric rare-earth oxides or oxide-fluorides, TiO_{2-x} and WO_{3-x} where the defects are arranged along particular directions or in parallel planes[2,7,8]. In TiO_{2-x} and WO_{3-x} which have a large range of non-stoichiometry, several phases with related structures are found as a function of x. Segregation of defects occurs in "shear planes"; the number of planes and the spacing between them can be directly deduced from the chemical composition [9-14]. When defect-ordering appears in two different directions, it leads to block structures as observed in $NaNbO_3$-$Ca_2Nb_2O_7$ or $NaTaO_2F_2$-TaO_2F systems[15-17].

Up to now, very few authors have dealt with vacancy ordering in ternary oxides AMO_{3-y} derived from the perovskite structure. Some have been prompted to study more carefully vacancy-ordering in some phases of this type and to correlate it to significant physical properties.

2 Previous Studies

Many perovskite compounds AMO_{3-y}(A = Ca, Sr, Ba; M = 3 d element) exhibit non-stoichiometry in a wide range, $O < y < 0.50$[18]. They have been investigated thoroughly because of the interest in the physical properties resulting from the presence of two oxidation states of the metallic cation. The most important results that have been obtained in this field and which are related to the influence of anionic defects are summarized in the following.

From a general point of view, the composition of these oxides depends both on the annealing temperature and the oxygen partial pressure maintained during the synthesis, these parameters determining the ratio of oxygen defects. This explains the various ways of synthesizing such phases[19]. Structural studies have been carried out using X-ray diffraction in order to determine the range of existence of the perovskite structure as a function of y[20]. This sometimes extends to $y = 0.50$ although it corresponds to a large percentage of vacancies ($\simeq 17\%$). Examples include $SrTiO_{2.5}$, $SrVO_{2.5}$ or the high-temperature phase of $SrFeO_{2.50}$[21–23]. Previous authors concluded that the defects were randomly distributed. However as the techniques of investigation have recently been improved, however, the possibility of vacancy-ordering has been considered. Using electron microscopy, Alario Franco observed superlattice fringes in $SrTiO_{2.50}$ implying a possible ordering of the vacancies in this phase[24]. Employing similar criteria, Tofield et al. have given evidence for the existence of a new ordered phase $SrFeO_{2.75}$ in the $SrFeO_{3-x}$system[25]. On the basis of superlattice reflections, they proposed a model for the structure of $SrFeO_{2.75}$ related to that of $SrFeO_{2.50}$, assuming also a vacancy-ordering along some [110] rows. Thus, half of the iron atoms remain sixfold coordinated while the other half becomes fivefold coordinated. Similar ordering patterns were established in the 4 H $(Ba_{0.50}Sr_{0.50})MnO_{2.84}$ and 6 H $BaFeO_{3-x}$ phases using neutron-diffraction data. In $Ba_{0.5}Sr_{0.5}MnO_{2.84}$ the oxygen vacancies were found to be in hexagonal (Ba, SrO_3) layers, suggesting that the Mn^{3+} ions are coordiated by oxygen in edge-sharing trigonal pyramids[26]. In the 6 H $BaFeO_{2.79}$ phase, the vacancies were found to be distributed in every fourth [110] row to give $(BaO_{2.50})$ hexagonal layers and $(BaO_{2.835})$ cubic layers. This may lead to one quarter of corner sharing $Fe^{III}O_4$ tetrahedra, one quarter of $Fe^{III}O_6$ octahedra and one half of $Fe^{IV}O_6$ octahedra [27]. As will be shown later, the great importance of the metal-oxygen coordination which seems to govern the variation of the structure with the stoichiometry cannot be over-emphasized. The main properties which have been studied up to now, are the magnetism, the electronic conductivity and the O^{2-} conductivity. These properties are related to the mixed valencies of M cations and/or the occurrence of anionic defects[28–35]. One system of particular interest is cobaltite for which a low spin-high spin transition occurs with increasing temperature, which leads to unusual behaviour and remarkable physical properties as y varies (electrocatalytic activity for instance[36]) [28, 37].

Of all the systems mentioned, ferrites seem to be the most extensively investigated phases, due to the fact that the environment and the valency of iron can be determined by Mössbauer spectroscopy. $AFeO_{3-y}$ (A = Sr, Ba) ferrites are very interesting materials for studying the influence of non-stoichiometry on local iron coordination by means of Mössbauer resonance[38, 39]. Besides the classical physical properties of these phases which have also been studied, these additional investigations have led some authors to propose hypotheses on the existence of vacancy ordering[40–42]. For instance, with the $La_{1-2y}Sr_{2y}FeO_{3-y}$ system, Yamamura and Kiryama concluded that iron exists both in octahedral and tetrahedral sites over the whole composition range ($0 < y \leqslant 0.50$). Moreover, for a small range, $0.30 \lesssim y \lesssim 0.45$, they found that iron was fivefold coordinated[43]. Such a coordination has also been supposed by Tofield et al. in $SrFeO_{2.75}$[25, 44]. Generally, although only a few systems have been carefully investigated, it seems that for low vacancy concentrations the symmetry of the perovskite remains unchanged but that important changes can occur in the structure as y increases. However, most of this work is rather limited in scope and no model of non-stoichiometry has really been established for these phases.

In the follwing we propose an original structural model for the AMO_{3-y} phases based on the relationship between the perovskite and brownmillerite structure types[45].

3 Hypothesis of the Structure of Non-Stoichiometric Perovskites AMO_{3-y}

First of all, we notice that a constant ratio of A to M cations (A/M = 1) prevents the formation of shear planes of the type observed in WO_{3-x} even if y varies considerably ($0 \leq y \leq 0.50$) (Fig. 1).

This finding shows that a model of non-stoichiometry for the AMO_{3-y} phases completely different from that of the shear mechanism is required. This model may be deduced from the brownmillerite-like structure of dicalcium ferrite $Ca_2Fe_2O_5$ (or $CaFeO_{2.50}$). This compound can be considered as a non-stoichiometric perovskite AMO_{3-y} (A = Ca; M = Fe) corresponding to y = 0.50 and its structure is easily derived from the cubic perovskite structure AMO_3, assuming that the oxygen vacancies (one per six anionic sites) are ordered along [101] rows in each second (0k0) plane. A slight shift of the iron atoms in those planes leads to a structure with alternating sheets of (MO_6) octahedra and (MO_4) tetrahedra perpendicular to the y axis (Fig. 2). This relationship between the two structures suggests a general model for the AMO_{3-y} phases also with intermediate compositions ($0 < y < 0.50$) also. It supposes a sequence of $(n-1)$ perovskite-like sheets of MO_6 octahedra alternating with one sheet consisting of parallel chains of (MO_4) tetrahedra. The general formula of such phases would be $A_nM_nO_{3n-1}$ ($n \geq 2$; $y = 1/n$). Their idealized structures for n = 2, 3, 4 and ∞ are shown in Fig. 3.

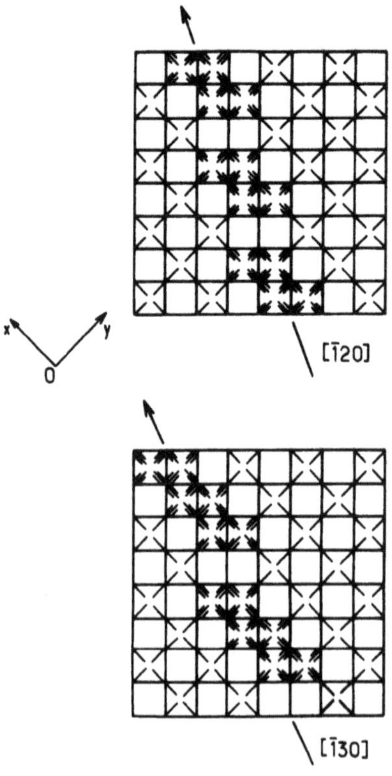

Fig. 1. Shear plane formation in the W_nO_{3n-1} (*above*) and W_nO_{3n-2} (*below*) systems[14]

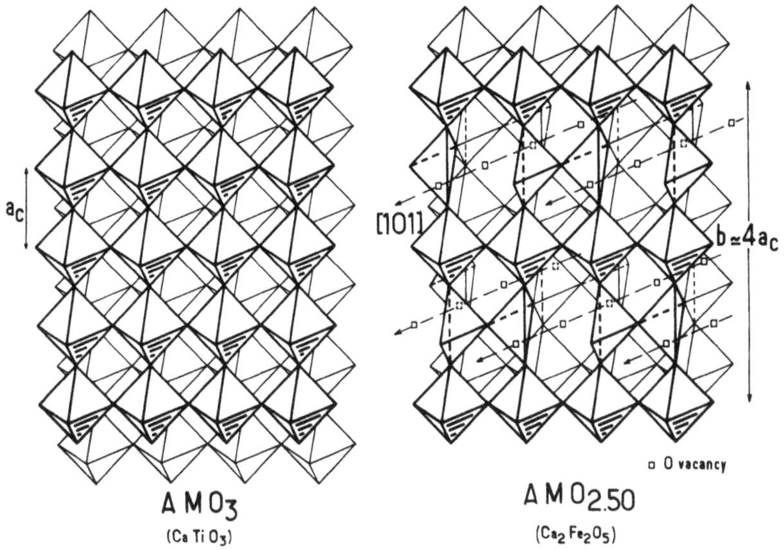

Fig. 2. Idealized structure of $CaTiO_3$ (perovskite) and $Ca_2Fe_2O_5$ (brownmillerite-like structure)

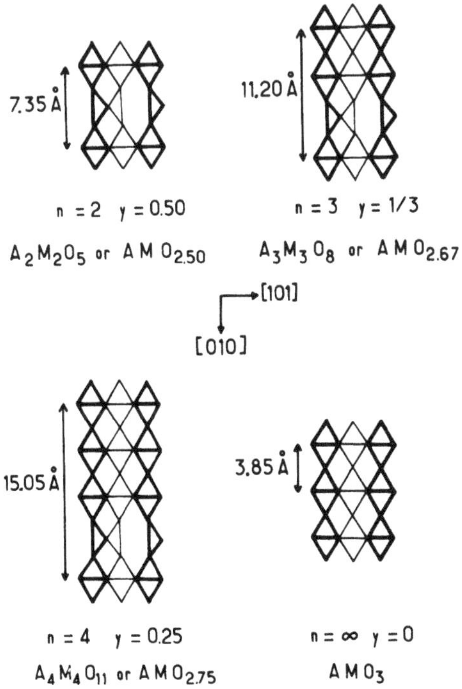

Fig. 3. Idealized structures of $A_nM_nO_{3n-1}$ phases

Assuming an orthorhombic symmetry (similar to that of $Ca_2Fe_2O_5$), the theoretical cell parameters are related to that of the perovskite a_c:

$$a_n \simeq \sqrt{2}\, a_c, \quad b_n \simeq n a_c, \quad c_n \simeq \sqrt{2}\, a_c$$

To check this model, several $A_2M_2O_5$-AMO_3 systems have been systematically investigated[45]. We will report here some significant results.

4 Non-Stoichiometry of Some AMO_{3-y} Systems

A complete study of three systems has been carried out using several techniques such as X-ray diffraction, electron microscopy, Mössbauer spectroscopy, magnetic measurements, and electrical conductivity. Two methods have been used to create anionic defects: – replacement of lanthanum by calcium or strontium in $La_{1-2y}Ca_{2y}Fe^{III}O_{3-y}$ and $La_{1-2y}Sr_{2y}Fe^{III}O_{3-y}$ – replacement of titanium by trivalent iron in $CaTi_{1-2y}Fe_{2y}O_{3-y}$.

4.1 Synthesis and X-Ray Characterization

Samples of various composition were prepared from stoichiometric mixtures of carbonates ($CaCO_3$, $SrCO_3$) and oxides (La_2O_3, Fe_2O_3). The starting materials were ground and heated in air at 1100 °C. The reaction was complete after annealing at 1350 °C for three days. To obtain more homogeneous samples, the starting materials were dissolved in dilute nitric acid. The nitrates were then decomposed at low temperature ($\simeq$ 800 °C) and finally fired at 1300 °C.

As the samples contained a small amount of tetravalent iron, a further annealing under low partial oxygen pressure was necessary in order to reduce Fe^{IV} to trivalent iron. Typical conditions were:

$La_{1-2y}Ca_{2y}FeO_{3-y}$ $\quad p_{O_2} = 10^{-7}$ atm
$La_{1-2y}Sr_{2y}FeO_{3-y}$ $\quad p_{O_2} = 10^{-12}$ atm
$CaTi_{1-2y}Fe_{2y}O_{3-y}$ $\quad p_{O_2} = 10^{-4}$ atm

X-ray diffraction analysis reveals two different vacancy distributions without the detection of a two-phase domain in every system. For low values of $y \lesssim 0.25$, the symmetry of the perovskite remains which indicates that the vacancies are apparently disordered. At higher values of $y \simeq 0.25$ X-ray patterns give evidence of a distortion. For $y \simeq 0.25$, they can be indexed with the theoretical parameters (orthorhombic symmetry) deduced from the vacancy ordering previously described. As a consequence, an order-disorder transition takes place around $y \approx 0.25$. This phenomenon is illustrated in Fig. 4, which shows a discontinuity of the unit cell volume V_m around this value for each system.

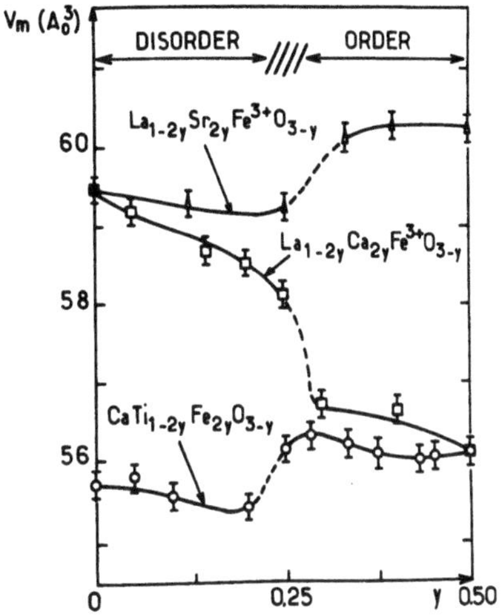

Fig. 4. Variation of the unit cell volume V_m with y for different AMO_{3-y} phases

4.2 Experimental Results of Electron Microscopy

As predicted by our model, the sequence of octahedral and tetrahedral layers should appear in electron microscopic images, assuming that their electronic densities differ sufficiently. We investigated more specificly the $CaTiO_3$-$Ca_2Fe_2O_5$ system, i.e. the $CaTi_{1-2y}Fe_{2y}O_{3-y}$ solid solution[46]. First of all, dicalcium ferrite $Ca_2Fe_2O_5$ (y = 0.5, n = 2) was examined. An image obtained perpendicular to the y axis shows regularly spaced fringes (Fig. 5 a). The measured spacing, which is about 7.35 Å, actually corresponds to the distance between two planes of defects in $Ca_2Fe_2O_5$. This result is confirmed by the electron diffraction pattern. After that the $CaTi_{0.33}Fe_{0.67}O_{2.67}$ (y = 0.33; n = 3, $Ca_3Fe_2TiO_8$) and $CaTi_{0.50}Fe_{0.50}O_{2.75}$ (y =0.25; n = 4, $Ca_4Fe_2Ti_2O_{11}$) phases have been

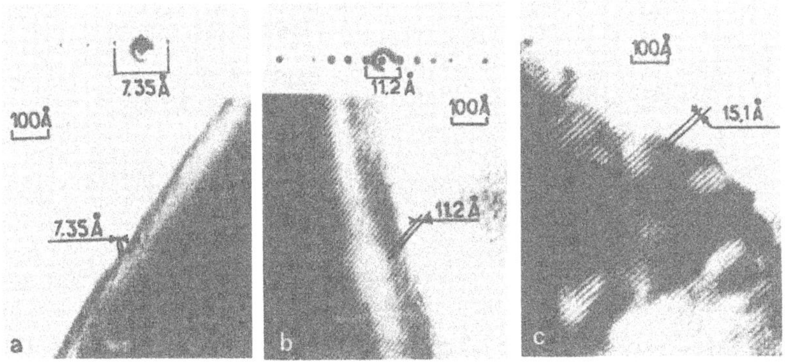

Fig. 5 a–c. Lattice images and electron diffraction patterns of $CaTi_{1-2y}Fe_{2y}O_{3-y}$ phases [y = 0.50 (**a**), 0.33 (**b**), 0.25 (**c**)]

investigated. Lattice images show the expected spacings between planes containing tetrahedra (Fig. 5b, c in correlation with Fig. 3). These results clearly verify our structural model, implying that for $y \geqslant 0.25$ the vacancies form rows giving rise to tetrahedral files parallely ordered in certain (0k0) planes. For intermediate compositions ($2 < n < 4$; $0.25 < y < 0.50$) we tried to interpret the continuous evolution of V_m with y observed by X-ray analysis (Fig. 4).

Microscopic examinations of $CaTi_{0.20}Fe_{0.80}O_{2.60}$ (or $Ca_5Fe_4TiO_{13}$ with $y = 0.40$; $n = 2.5$) (Fig. 6) reveal a crystal structure which is almost ordered. However, instead of

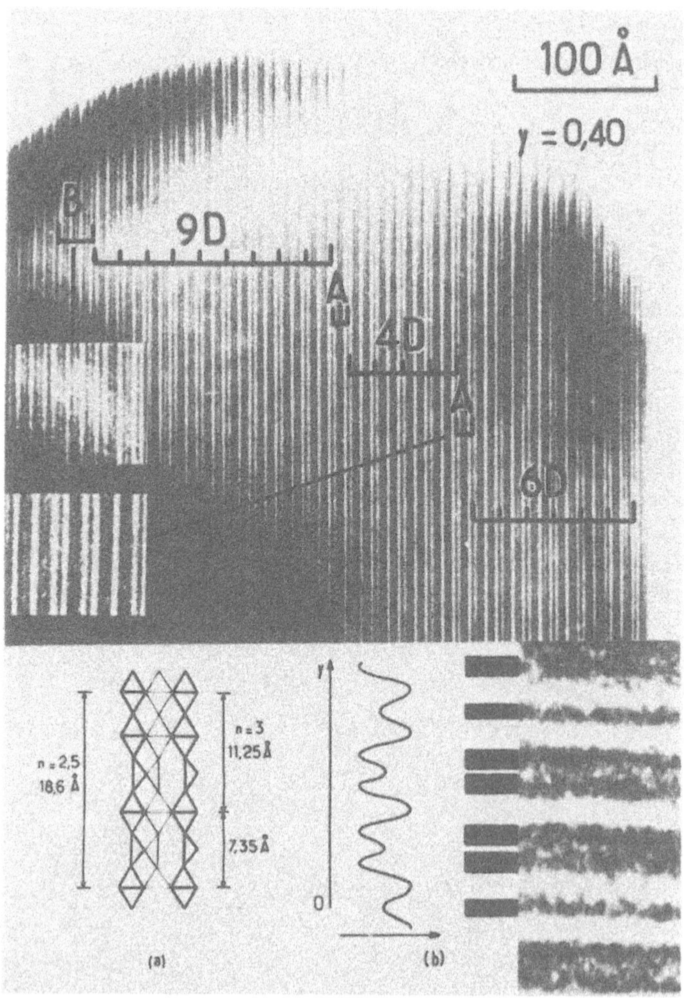

Fig. 6a, b. Lattice image of $CaTi_{0.2}Fe_{0.8}O_{2.60}$ ($n = 2.50$; $y = 0.40$).
a Idealized structure of $CaTi_{0.2}Fe_{0.8}O_{2.60}$ ($n = 2.50$);
b schematic variation of the electronic density along Oy and of the corresponding fringes

equidistant fringes, a more complicated image with a period of 18.6 Å (D spacing in Fig. 6) appears. This can be correlated to the hypothetic structure of the compound which would correspond to an ordered sequence of $n = 2$ and $n = 3$ terms (Fig. 6a) and to the schematic variation of the electronic density as shown in Fig. 6b. Two kinds of structural defects have been observed: the A type corresponds to the absence of an $n = 2$ term whereas the B type is the consequence of an additional $n = 2$ term. It can reasonably be concluded that the initial composition is maintained by a statistical distribution of these defects in the crystal.

Similar periodical fringe arrangements occur in other intermediary phases. Nevertheless, a series of intergrowths corresponding to n values close to definite compositions have been observed. Two examples may be taken as typical of this phenomenon:

- for $CaTi_{0.14}Fe_{0.86}O_{2.57}$ (or $Ca_7Fe_6TiO_{18}$ with $y = 0.43$; $n = 2.33$) three different kinds of microdomains were identified by their spacings: $D \simeq 18.6$ Å ($n = 2.50$), $E \simeq 26$ Å ($n = 2.33$) and $F \simeq 33$ Å ($n = 2.25$) (Fig. 7).
- for $CaTi_{0.10}Fe_{0.90}O_{2.55}$ (or $Ca_{11}Fe_{10}TiO_{28}$ with $y = 0.45$; $n = 2.20$) two large microdomains are identifiable corresponding to $n = 2$ and $n = 2.50$ (Fig. 8).

A similar observation has been made for $Ca_7Ti_3Fe_4O_{19}$ ($n = 3.5$; $y = 0.29$) and $Ca_8Ti_2Fe_6O_{21}$ ($n = 2.67$; $y = 0.37$). Thus, it can be concluded that for all these compositions ($y \geqslant 0.25$; $2 \leq n \leq 4$) an increase of n (or a decrease of y) is directly connected to a gradual addition of octahedral sheets perpendicular to Oy starting from the initial $Ca_2Fe_2O_5$ sequence with $n = 2$. This explains the continuous variation of V_m (Fig. 4). So far, there are relatively few octahedral planes present and a tendency towards long-range ordering may be detected.

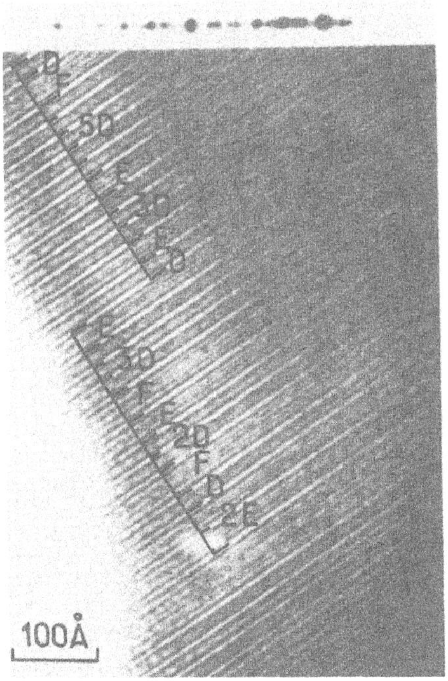

Fig. 7. Lattice image of $CaTi_{0.14}Fe_{0.86}O_{2.57}$ ($n = 2.33$; $y = 0.43$) with $D \simeq 18.6$ Å, $E \simeq 26$ Å and $F \simeq 33$ Å

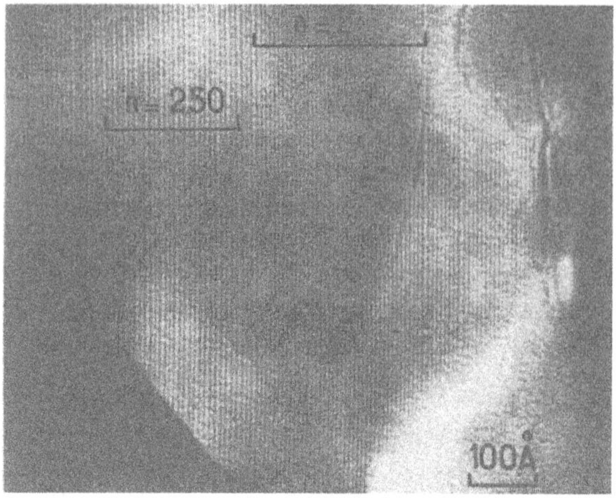

Fig. 8. Lattice image of $CaTi_{0.10}Fe_{0.90}O_{2.45}$ (n = 2.20; y = 0.45)

For lower vacancy ratios (y < 0.25; n > 4), electron diffraction patterns give evidence of a cubic symmetry. No ordering along [101] rows in (0k0) planes seems to take place, which is consistent with X-ray results. However, for $CaTi_{0.6}Fe_{0.4}O_{2.80}$ (y = 0.20; n = 5) some domains show a beginning of short-range ordering (Fig. 9). Microdomains

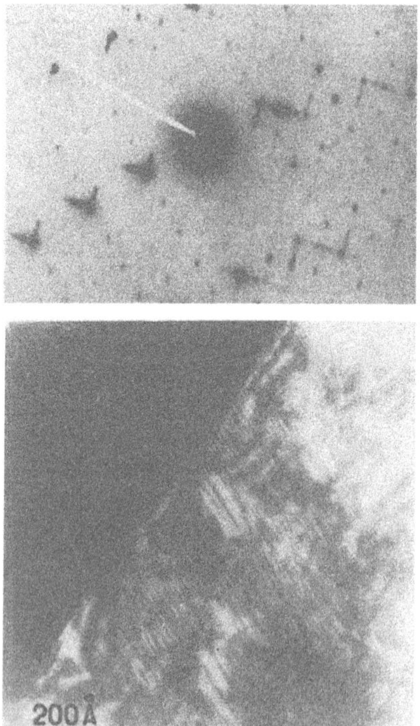

Fig. 9. Electron diffraction pattern and lattice image of $CaTi_{0.6}Fe_{0.4}O_{2.80}$ (n = 5; y = 0.20)

of fringes with rather large spacings can be observed in perpendicular directions. This corresponds to a cubic diffraction pattern with additional small reflections. Nevertheless, additional information was needed. Mössbauer spectroscopy seemed to be a complementary and appropriate technique of determining the coordination of iron and consequently the location of the vacancies for low y values as well as for $0.25 < y < 0.50$ where the symmetry of the fourfold coordination was hitherto unknown.

4.3 Mössbauer Resonance Studies

The $La_{1-2y}Ca_{2y}FeO_{3-y}$ solid solution containing only Fe^{3+} and the $CaTi_{1-2y}Fe_{2y}O_{3-y}$ solid solution containing simultaneously Ti^{4+} and Fe^{3+} have been systematically investigated[47, 48]. The Mössbauer resonance spectra were obtained using a classical device. Refinement of the spectra revealed the coordination number (C. N.) of iron. Assuming that the recoil-free fraction is the same for iron in each site, the ratio of iron in these sites has been deduced from the peak areas. We will first discuss the data on the $CaTi_{1-2y}Fe_{2y}O_{3-y}$ phases. The Mössbauer parameters for various compositions are given in Table 1.

4.3.1 The $CaTi_{1-2y}Fe_{2y}O_{3-y}$ System

Ordered structures: $0.25 \leqslant y \leqslant 0.50$

$y = 0.50$: $CaFeO_{2.50}$. The Mössbauer spectrum of the antiferromagnetic dicalcium ferrite is well known ($T_N = 275$ K)[49]. At room temperature, it reveals two Zeeman patterns corresponding to octahedral and tetrahedral Fe^{3+} ion sites as shown in Fig. 10. The corresponding parameters are reported in Table 1.

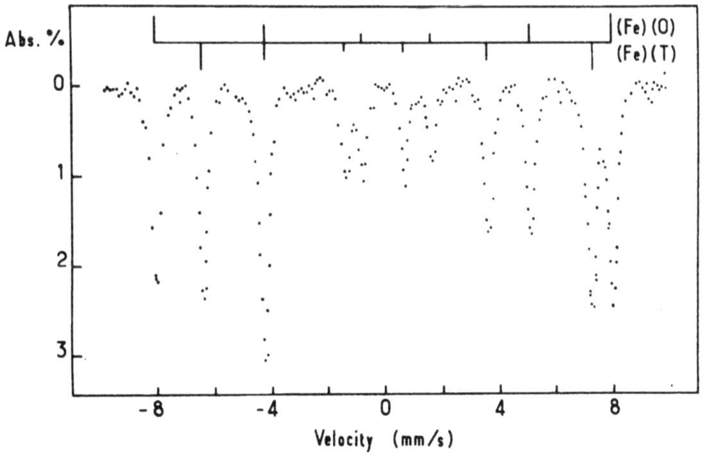

Fig. 10. Mössbauer resonance spectrum of $Ca_2Fe_2O_5$ at 295 K

Table 1. Mössbauer data of the $CaTi_{1-2y}Fe_{2y}O_{3-y}$ phases for various compositions

y	Phases	δ_O ($\pm$0.02 mm/s)	δ_T ($\pm$0.02 mm/s)	H_O ($\pm$5 kOe)	H_T ($\pm$5 kOe)	Δ_O ($\pm$0.03 mm/s)	Δ_T ($\pm$0.03 mm/s)	Fe(O)/Fe(T)
0.50	$Ca_2Fe_2O_5$							
	(4.2 K)			550	480	1.40	1.38	1.0 ± 0.02
	(295 K)	0.37	0.21	509	429	–	–	
0.40	$CaTi_{0.20}Fe_{0.80}O_{2.60}$	δ_{O_1} δ_{O_2}		H_{O_1} H_{O_2}				
	(4.2 K)	0.47 0.48	0.31	530 514	453	–	–	1.0 ± 0.02
0.33	$CaTi_{0.33}Fe_{0.67}O_{2.67}$							
	(4.2 K)	0.51	0.31	506	436	–	–	1.0 ± 0.1
	(345 K)	0.29	0.14	–	–	0.69	1.36	
0.25	$CaTi_{0.50}Fe_{0.50}O_{2.75}$							
	(4.2 K)	0.50	0.31	490	420	–	–	1.1 ± 0.3
	(295 K)	0.34	0.17	–	–	0.66	1.42	
0.20	$CaTi_{0.60}Fe_{0.40}O_{2.80}$							
	(4.2 K)	0.50	0.32	478	405	–	–	1.3 ± 0.4
	(295 K)	0.34	0.16	–	–	0.64	1.45	
0.15	$CaTi_{0.70}Fe_{0.30}O_{2.85}$							
	(295 K)	0.33	0.16	–	–	0.63	1.47	2.0 ± 0.5
0.10	$CaTi_{0.80}Fe_{0.20}O_{2.90}$							
	(295 K)	0.34	0.16	–	–	0.55	1.45	2.8 ± 0.7
0.05	$CaTi_{0.90}Fe_{0.10}O_{2.95}$							
	(295 K)	0.33	0.16	–	–	0.58	1.43	3.2 ± 1.0
0.01	$CaTi_{0.98}Fe_{0.02}O_{2.99}$							
	(295 K)	0.32	0.19	–	–	0.64	1.20	8.6 ± 1.5

δ_O, δ_T: isomer shifts; H_O, H_T: hyperfine fields; Δ_O, Δ_T: quadrupole splittings; Fe(O)/Fe(T): ratio between Fe in octahedral and tetrahedral site

$y = 0.40$: $CaTi_{0.20}Fe_{0.80}O_{2.60}$. The microscopic images obtained for this composition show an ordered sequence of octahedral and tetrahedral sheets with a ratio 3:2 (see Fig. 6). The 4.2 K Mössbauer spectrum indicates magnetic ordering (Fig. 11a). Magnetic susceptibility measurements have shown that this phase is antiferromagnetic with $T_N = 543$ K. The two Zeeman Patterns actually correspond to sixfold and fourfold coordinated Fe^{3+} ions. From the ratio of the peak areas (Fe(O)/Fe(T) = 1.0) it can be concluded that the oxygen tetrahedra are exclusively filled with iron, while titanium and iron are randomly distributed (1/3 and 2/3) over the octahedral sites. Thus, the cationic distribution can be schematically written: $Ca[Fe_{0.40}Ti_{0.20}]_O[Fe_{0.40}]_TO_{2.60}$. In addition, the lines corresponding to the octahedral sites are split into two. The refinement of the spectrum leads to a $Fe(O_2)/Fe(O_1)$ ratio between these sites close to 1.9. In view of the structure this result is not surprising: it is possible to distinguish in fact two types of octahedral sites with different environments (Fig. 11b). The O_1 site is surrounded by 4 octahedra containing statistically 2/3 Fe^{3+} and 1/3 Ti^{4+} and 2 tetrahedra containing exclusively Fe^{3+} while the O_2 site is surrounded by 5 octahedra with 2/3 Fe^{3+} and 1/3 Ti^{4+} and 1 tetrahedron with only Fe^{3+}. Therefore, the number of magnetic nearest neighbours is slightly higher for the O_1 site ($Z \simeq 4.67$) than for the O_2 site ($Z \simeq 4.33$). In agreement with previous results, we must ascribe the higher hyperfine field to the O_1 site[50)]. This explanation also confirms the ratio of 1.9 of the peak areas as there are twice as many O_2 than O_1 sites (Fig. 11c). The line width corresponding to the O_1 sites appears also to be smaller than that of the O_2 sites ($\Gamma_{O_1} \simeq 0.25$ mm/s; $\Gamma_{O_2} \simeq 0.50$ mm/s). This can be related to the stronger statistical effect of the cationic distribution around the O_2 octahedral sites.

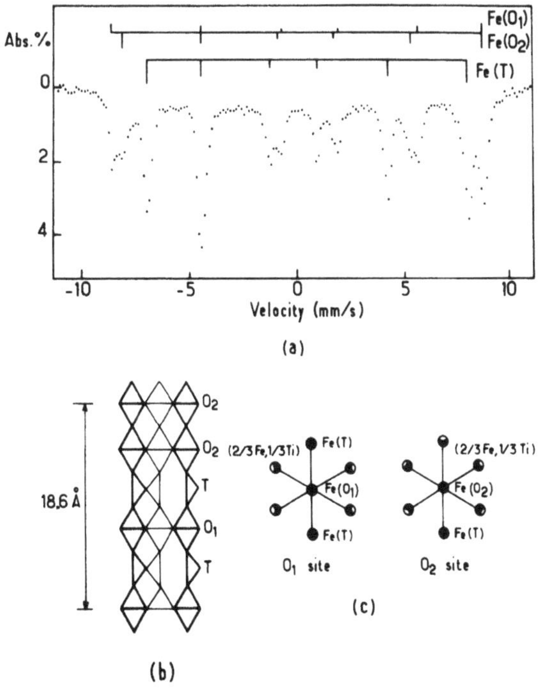

Fig. 11. **a** Mössbauer resonance spectrum of $CaTi_{0.2}Fe_{0.8}O_{2.60}$ at 4.2 K. **b** Idealized structure of this phase, **c** cationic environment of the O_1 and O_2 sites

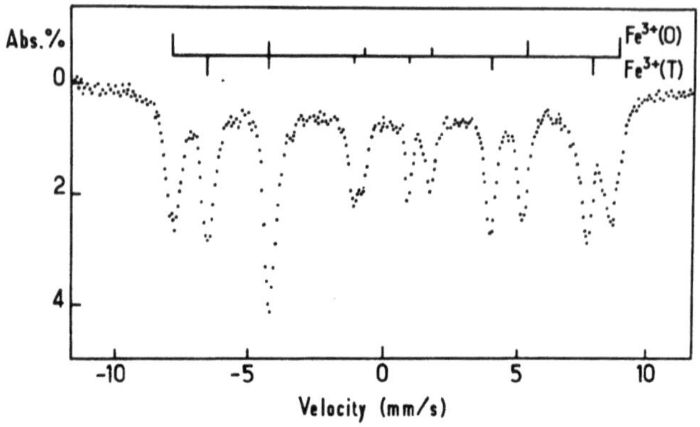

Fig. 12. Mössbauer resonance spectrum of $Ca_3Fe_2TiO_8$ (y = 0.33) at 4.2 K

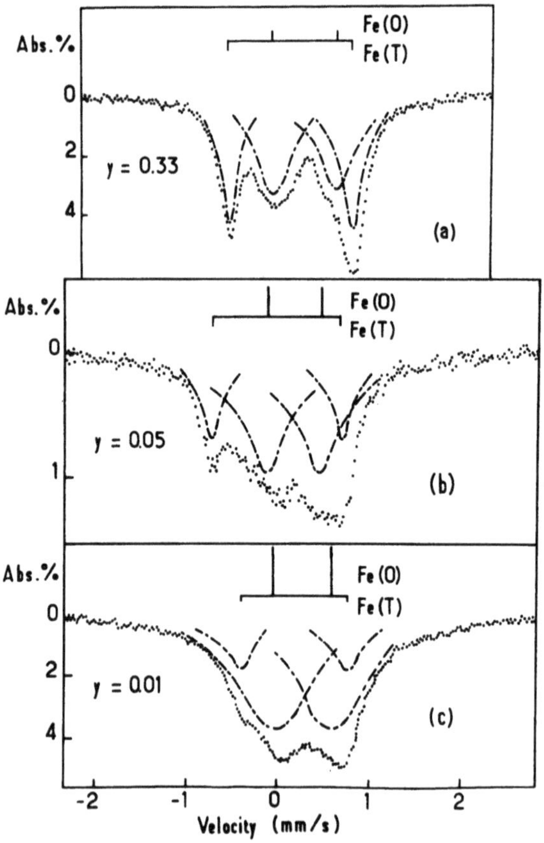

Fig. 13a–c. Mössbauer resonance spectra of $CaTi_{1-2y}Fe_{2y}O_{3-y}$ phases for y = 0.33, 0.05 and 0.01

$y = 0.33$: $CaTi_{0.33}Fe_{0.67}O_{2.67}(Ca_3Fe_2TiO_8)$. The Mössbauer resonance patterns of this phase have been recorded at 4.2 and 345 K. The 4.2 K spectrum – below the antiferromagnetic temperature ($T_N \simeq 305$ K) – is characteristic of trivalent iron, that is octahedrally and tetrahedrally coordinated, respectively (Fig. 12). The ratio of the peak areas corresponding to each site is 1.0 (Table 1). From the idealized structure (Fig. 3), this ratio leads to the cationic distribution:

$$Ca[Fe_{1/3}Ti_{1/3}]_O[Fe_{1/3}]_TO_{2.67}$$

Above the Néel temperature, the paramagnetic spectrum shows two doublets corresponding to both sites and the ratio Fe(O)/Fe(T) is also close to 1.0 (Fig. 13a).

$y = 0.25$: $CaTi_{0.50}Fe_{0.50}O_{2.75}$. According to the calculated parameters (Table 1), the Mössbauer spectra are similar to those obtained for y = 0.33. In our structural model (Fig. 3) the ratio of octahedral to tetrahedral layers is 3:1. The cationic distribution, as deduced from the ratio Fe(O)/Fe(T), is slightly higher than unity (Table 1) and the corresponding compound can be written as: $Ca[Fe_{0.25}Ti_{0.50}]_O[Fe_{0.25}]_TO_{2.75}$. This confirms the ordered structure observed in the microscopic images for this composition.

Usually, the fit of paramagnetic spectra is slightly inaccurate because of the non-Lorentzian profile of the widened octahedral peaks. This effect can be due to the existence of different octahedral sites or, more likely, to the random occupancy of these sites by Ti and Fe atoms. Due to the fact that the Fe(O)/Fe(T) ratio remains constant (= 1) in this range of composition, we can conclude that Fe^{3+} is equally distributed among the octahedral and tetrahedral sites. According to our structural model Ti^{4+} occupies octahedral sites exclusively.

Disordered Structures: $0 < y < 0.25$

In this section we will assume that the latter conclusion can be extended to the whole composition range, namely that the Ti^{4+} ions are never fourfold coordinated, even at high concentrations (low y). This hypothesis seems quite reasonable, because Ti^{4+} is very rarely found in tetrahedral coordination in oxides (the only known oxides with C. N. 4 are strongly basic titanates such as Na_4TiO_4, K_4TiO_4 or Ba_2TiO_4[51, 52]).

The investigation of the $CaTi_{1-2y}Fe_{2y}O_{3-y}$ solid solution for $y < 0.25$ by X-ray diffraction and electron microscopy has shown that the long-range symmetry appears to be that of perovskite, which implies a disorder of the anionic vacancies in the oxygen lattice. At room temperature, these phases are paramagnetic and the Mössbauer spectra can be refined with two doublets for each composition (see Figs. 13b, c). The isomer shifts are perfectly constant and characterize both octahedral and tetrahedral sites ($\delta_O \simeq 0.33$ mm/s, $\delta_T \simeq 0.17$ mm/s). The peaks corresponding to tetrahedral iron are very narrow for the whole composition range, typically 0.25–0.30 mm/s, whereas those related to octahedral sites are increasingly broadened from 0.50 mm/s for $y \simeq 0.25$ to 0.75 mm/s for $y = 0.01$. So far, no satisfactory explanation for this broadening may be proposed. Moreover the Fe(O)/Fe(T) ratio increases continuously as y decreases (Table 1). So it must be pointed out that even for very low iron concentrations (for instance y = 0.01), the existence of (FeO_4) tetrahedra may be still detected.

4.3.2 The $La_{1-2y}Ca_{2y}FeO_{3-y}$ System

As will be seen below, the Néel temperature of these phases is very high ($T_N > 700$ K). The Mössbauer spectra obtained at room temperature show two Zeeman patterns, once again corresponding to trivalent iron in octahedral and tetrahedral sites. Mössbauer data are reported in Table 2 and conclusions similar to those for the $CaTi_{1-2y}Fe_{2y}O_{3-y}$ can be drawn. For ordered structures (for instance y = 0.40 and y = 0.33, Fig. 14), the Fe(O)/Fe(T) ratio is in perfect agreement with the ratio deduced from the predicted structures.

Table 2. Mössbauer data of $La_{1-2y}Ca_{2y}FeO_{3-y}$ phases

Y	δ_O (mm/s) ± 0.02	δ_T (mm/s) ± 0.02	H_O (T)	H_T (T)	Fe(O)/Fe(T)
0.05	0.39	–	53.1	–	–
0.10	0.39	0.22	52.3	41.0	12 ± 0.5
0.25	0.39 0.37	0.225	53.9 52.2	40.1	3.6 ± 0.3
0.33	0.375	0.19	51.0	42.6	1.99 ± 0.05
0.40	0.39 0.37	0.21	52.7 51.9	43.1	1.47 ± 0.05
0.50	0.37	0.21	50.9	42.9	0.98 ± 0.02

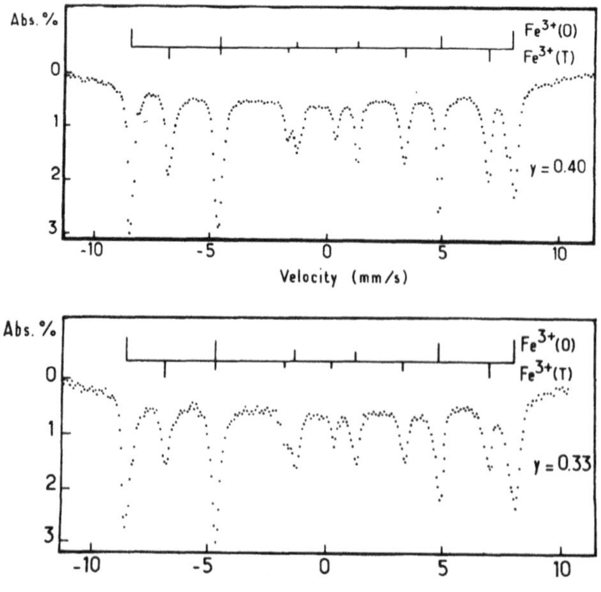

Fig. 14. Mössbauer resonance spectra of $La_{1-2y}Ca_{2y}FeO_{3-y}$ phases (y = 0.40 and y = 0.33) at 295 K

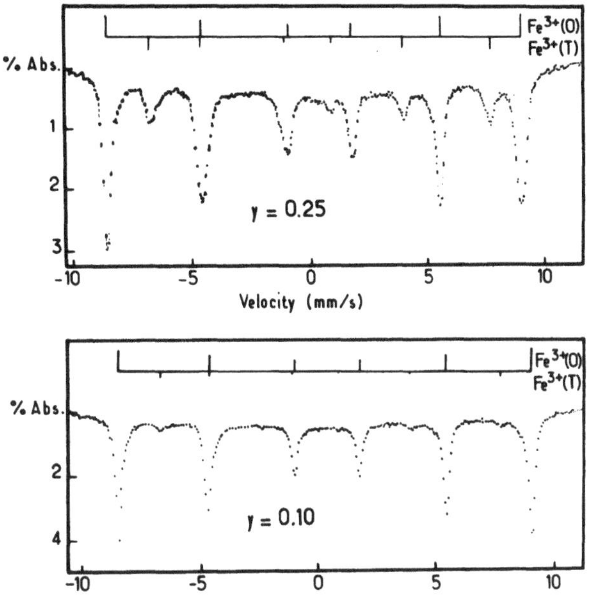

Fig. 15. Mössbauer resonance spectra of $La_{1-2y}Ca_{2y}FeO_{3-y}$ phases (y = 0.25 and y = 0.10) at 295 K

This once again is validifies by our model. As y decreases (y < 0.25), this ratio is no longer equal to that corresponding to an ideal long-range ordering. For lower y values, it becomes difficult to detect small quantities of Fe in tetrahedral sites (Fig. 15). For y = 0.05, no peaks corresponding to Fe in such a site appear. Nevertheless, as in the previous study, no fivefold coordinated iron was found.

4.3.3 Conclusions

From the cited results it may be concluded that the Mössbauer parameters (isomer shift, hyperfine field and quadrupole splitting; Tables 1 and 2) vary continuously with the composition of the system. The observed isomer shifts give evidence of the presence of trivalent iron in tetrahedral sites for the whole composition range and even for very low iron concentrations. Using the area ratio Fe(O)/Fe(T) denoted by R, the cationic distribution in these phases could be determined. Moreover, it is possible to calculate the molar fraction of tetrahedra, t(y), assuming that only Fe^{3+} ions are located in these sites. For the $CaTi_{1-2y}Fe_{2y}O_{3-y}$ and $La_{1-2y}Ca_{2y}FeO_{3-y}$ solid solutions, t(y) is given by the relations $t(y) = \frac{2y}{R+1}$ and $t(y) = \frac{1}{R+1}$, respectively. Using the experimental data of Tables 1 and 2, the variation of t(y) with y is reported in Fig. 16. It is significant that a similar change is observable for both systems for the whole range of composition. The rather continuous order-disorder transition appears close to y = 0.25. For y > 0.25, the linear variation of t(y) with y confirms the structural model and thus the previous results, i.e. the presence of alternating octahedral and tetrahedral layers in the structure.

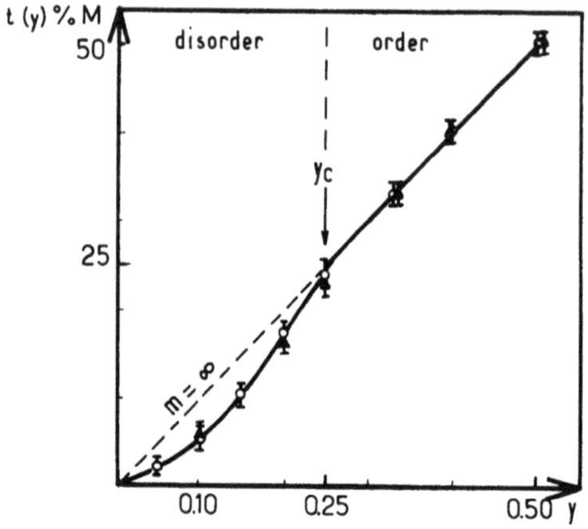

Fig. 16. Experimental variation of the molar fraction the tetrahedra vs. y for different AMO_{3-y} phases

For smaller values of $y \lesssim 0.25$, despite the deviation from ($m = \infty$), the existence of tetrahedra gives evidence of short-range order. Thus, an unexpected tendency of the anionic defects to couple in the neighbourhood of ions such as Fe^{3+}, inducing a change from C. N. 6 to 4 is found. A model taking account of this unusual segregation phenomenon is proposed in Chap. 7.

The above results concerning the behaviour of non-stoichiometric AMO_{3-y} perovskites are significant in that, for the first time, evidence of short-range ordering has been found for low-vacancy concentrations.

4.4 Magnetic Properties

As the magnetic properties of dicalcium ferrite $Ca_2Fe_2O_5$ are already known, it was considered interesting to investigate the magnetic behaviour of these solid solutions and to correlate it with the vacancy order-disorder transition. Thermomagnetic analyses were carried out to determine the Néel temperatures of these phases.

The $CaTi_{1-2y}Fe_{2y}O_{3-y}$ phases are antiferromagnetic at low temperature when y is greater than a critical value $y \simeq 0.125$. The Néel temperature T_N decreases strongly with y (Fig. 17)[53]. For $y < 0.125$, a paramagnetic behaviour is observed for the whole temperature range considered. The variation of T_N with y is characteristic of a magnetic system with Fe^{3+} ions progressively diluted by diamagnetic ions. Taking into account the antiferromagnetic behaviour observed at low temperature, it is reasonable to conclude that $CaTi_{1-2y}Fe_{2y}O_{3-y}$ phases have a G-type magnetic structure which is similar to those for $LnFeO_3$ orthoferrite and $Ca_2Fe_2O_5$ ferrite[54, 55].

On the other hand, the $La_{1-2y}Ca_{2y}FeO_{3-y}$ phases are antiferromagnetic over the whole range of composition. The Néel temperature decreases only slightly from 750 K for $LaFeO_3$ to 725 K for $CaFeO_{2.50}$ (Fig. 17).

The order-disorder vacancy transition close to $y = 0.25$ does apparently not affect the magnetic behaviour, probably because the magnetic coupling depends essentially on the number of nearest neighbours of Fe^{3+}. This number is simply a function of y for the first solid solution, and a constant in the second case.

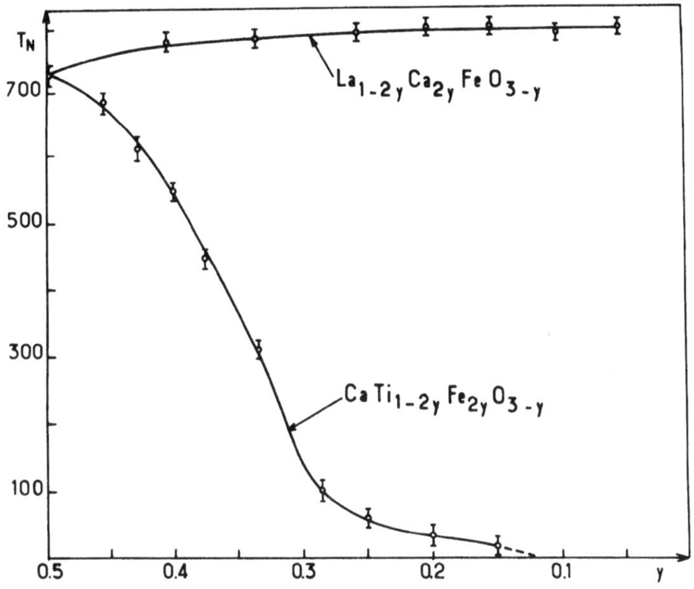

Fig. 17. Variation of the Néel temperatures of $CaTi_{1-2y}Fe_{2y}O_{3-y}$ and $La_{1-2y}Ca_{2y}FeO_{3-y}$ solid solutions with y

5 Some Other Ordered Non-Stoichiometric Perovskites

Although several analogous systems have been investigated in the past, no phase belonging to our structural model has been reported[18]. Nevertheless, we were able to synthesize and characterize two new series corresponding to n = 3 terms, namely $Ca_2AFe_3O_8$ (A = Y, La, Gd) and $Ca_2AFe_2TiO_8$ (A = Ca, Sr, Ba). These phases have been prepared under controlled partial oxygen pressure to prevent the formation of tetravalent iron. X-ray patterns were indexed on the basis of theoretical cell parameters (orthorhombic symmetry). The results are compiled in Table 3. Attempts to grow single crystals in order to determine the precise structures were unsuccessful. At high temperature, crystallographic transitions occur. For example, the $Ca_2LaFe_3O_8$ (y = 0.33) ferrite has a cubic perovskite-type structure above 1350 °C (a = 3.868 ± 0.002 Å). The chemical analysis indicates an unpredicted excess of oxygen. This transition can be schematically described by:

$$Ca_2LaFe_3O_8 + x/2\ O_2 \leftrightarrow Ca_2LaFe^{3+}_{3-2x}Fe^{4+}_{2x}O_{8+x}$$

Orthorhombic phase cubic phase

Typical experimental runs gave the following results. After an annealing period of two days at 1400 °C in air, x is equal to 0.125, implying the presence of 8.3% Fe^{4+} [approximate composition: $(Ca_{2/3}La_{1/3})Fe^{3+}_{0.92}Fe^{4+}_{0.08}O_{2.71}$]. The crystallographic transition is indicative of a long range vacancy disorder in the high-temperature phases.

Table 3. Cell parameters of some $A_3M_3O_8$ phases

Phases	a (± 0.003 Å)	b (± 0.010 Å)	c (± 0.003 Å)
$Ca_3Fe_2TiO_8$	5.444	11.210	5.532
$Ca_2SrFe_2TiO_8$	5.462	11.311	5.570
$Ca_2BaFe_2TiO_8$	5.504	11.385	5.597
$Ca_2YFe_3O_8$	5.456	11.226	5.545
$Ca_2LaFe_3O_8$	5.464	11.293	5.563
$Ca_2GdFe_3O_8$	5.420	11.288	5.500

The non-stoichiometric $Ca_2LaFe_3O_{8+x}$ ferrite was also studied by measuring the electrical conductivity variation with oxygen partial pressure, p_{O_2}, at various temperatures (Fig. 18). In each curve, three ranges of variation are observable as represented for the imaginary temperature t_i:

AB range: $p_{O_2} < 10^{-10}$ atm

For very low oxygen pressures, the small variation of σ implies that the sample has almost a stoichiometric composition and contains nearly equal amounts of n and p carriers.

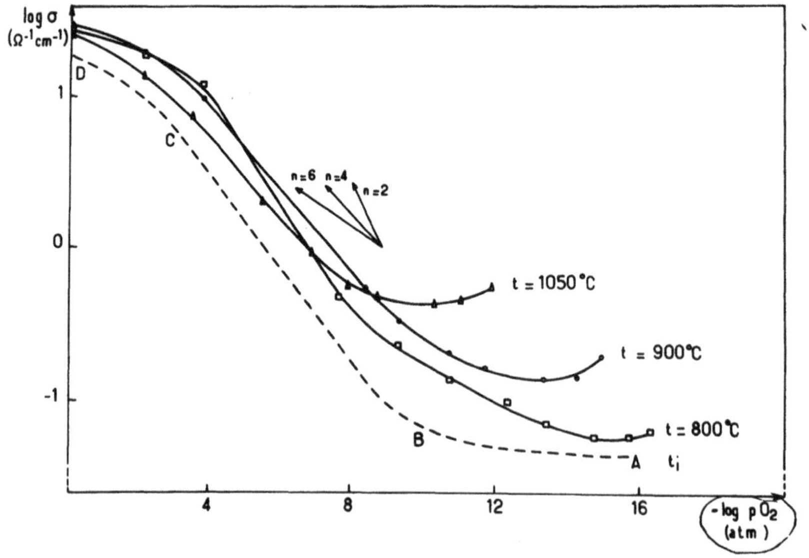

Fig. 18. Variation of the electrical conductivity [log σ] of $Ca_2LaFe_3O_{8+y}$ with the oxygen partial pressure at various temperatures

BC range: $10^{-10} < p_{O_2} < 10^{-4}$ atm

The electrical conductivity σ strongly increases with p_{O_2}. The shape of the curve characterizes a p semi-conductor. The conductivity approximately follows a law of the type:

$$\sigma = k\, p_{O_2}^{1/n_T}$$

From the linear variation of log σ with p_{O_2} and from the calculated values of n_T ($2.78 \leq n_T \leq 3.84$) for $800 < T < 1050\,°C$, a mechanism to explain the crystallographic transition has been proposed[56]. It implies that the non-stoichiometry is the consequence of the introduction of singly charged interstitial oxygen O_i^- into vacant lattice sites (Fig. 19) with a tendency to form $(O_i)_2^{2-}$ pairs. As a consequence, the latter modify the tetrahedral environment of some of the Fe^{3+} ions, which become octahedral. The increase of the number of these octahedra with the excess of oxygen finally leads to a structural transition. The ordered sequence of octahedral and tetrahedral layers is broken and a long range cubic symmetry of the perovskite appears.

CD range; $10^{-4} < p_{O_2} < 1$ atm

The non-linear variation of log σ is probably caused by a complex conduction mechanism, due to the presence of other kinds of defects.

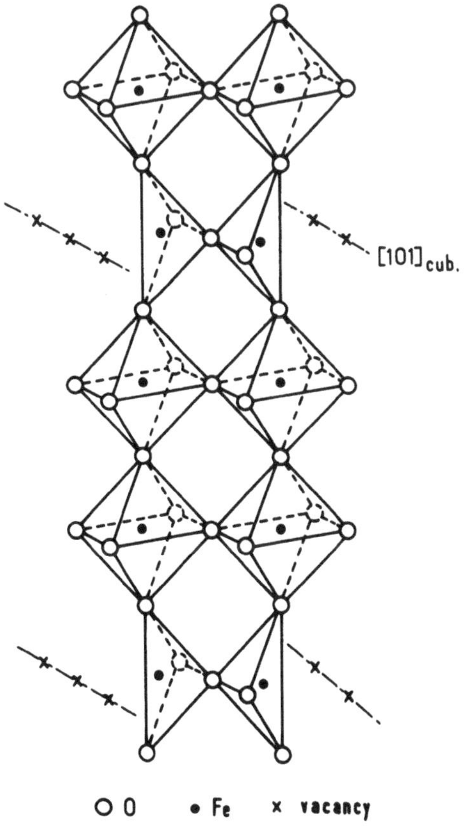

Fig. 19. Idealized structure of $Ca_2LaFe_3O_8$ showing vacancy strings

6 Vacancy Distribution in AMO_{3-y} Phases

The discussed results and in particular the Mössbauer data show that the number of tetrahedra per mole seems to be independent of the nature of the A and M cations. On the other hand a strong dependence on the ratio of anionic defects (y) is found. It would thus be interesting to find the vacancy distribution law and to correlate it with the molar percentage of tetrahedra t(y) in the AMO_{3-y} phases. A preliminary calculation was performed assuming a statistical distribution of the vacancies in the anionic network. The additional condition was that no more than two vacancies were present in the same octahedron. The counting of the defect clusters led to calculated values of t(y) [t(y)(1) in Fig. 20; for details of the calculations see Ref. 57]. The striking disagreement between the experimental and theoretical data demonstrates that an hypothesis, which only takes the metal coordination number into account, is not realistic. A second calculation was thus carried out assuming the C. N. of A to be [11], or [10] as in [12] as in the perovskite, or either brownmillerite structures, on the one hand and the C. N. of M to be 6, 5 or 4 on the other hand. On this basis, it was shown that the only possibility of arranging the vacancies is along rows of various length which are statistically distributed in one of the six directions of the perovskite network (Fig. 21). The theoretical curve for t(y) [t(y)(2) Fig. 20] is now in perfect agreement with the experimental data. This result appears to be very important in connection with the non-stoichiometry of perovskites. It allows a plausible model to be proposed.

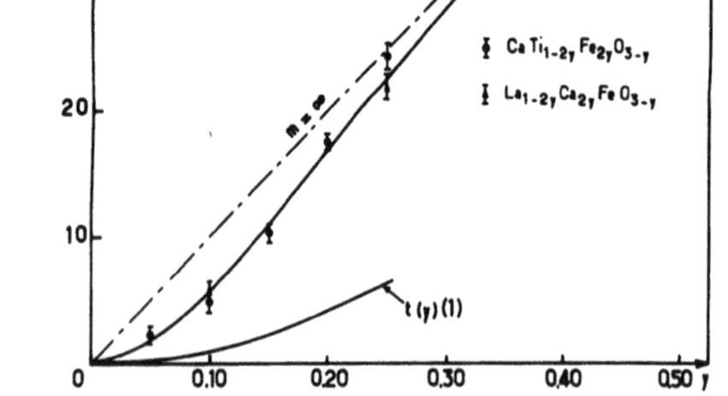

Fig. 20. Comparison between Mössbauer data and theoretical curves [experimental points ●, ▲; theoretical curves: t(z)(1), t(z)(2) – see text]

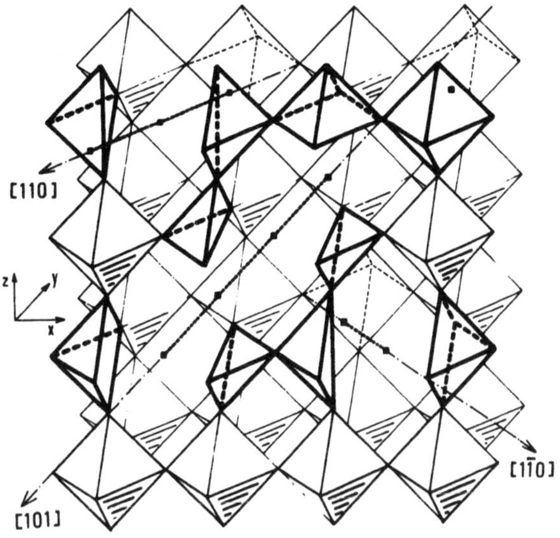

Fig. 21. Linear clusters of m vacancies in the perovskite lattice

7 Model of Non-Stoichiometry for AMO_{3-y} Phases

Based on the latter calculation and investigations of some AMO_{3-y} systems, a model which involves three steps for the vacancy ordering can be imagined:

– for low vacancy ratio, typically $y \lesssim 0.15$, the oxygen vacancies are ordered along rows of various length. Due to the fact that these rows are statistically oriented over the six possible directions within the network, the perovskite symmetry remains intact.

– as y increases close to $y \simeq 0.20$, the number of rows of infinite length increases rapidly. They become ordered parallel to each other in microdomains (see Fig. 21) which are randomly distributed in the lattice

– for $y \simeq 0.25$ (i.e. 8% vacancies) the number of these infinite rows is sufficiently large to induce long-range ordering. This corresponds to a parallel alignment of these files of defects in certain planes.

For $y > 0.25$, the increase of the vacancy concentration merely leads to a progressive addition of defect planes, i.e. to an increase in the number of tetrahedral layers with respect to the number of octahedral planes.

This model does not cover energetic aspects, especially at the disorder-order transition. Only electrostatic considerations concerning the cation coordinations have been examined, which explains the slight differences observed for $y > 0.25$. Moreover, our assumption of the "range of disorder" seems reasonable since it describes the observed phenomena. Finally, one understands better why, when the long-range ordering disappears close to $y = 0.25$, the number of tetrahedra is smaller than it would be if the structure was still ordered ($m = \infty$, Fig. 16).

In conclusion, we cannot expect that our model be extended to all AMO_{3-y} systems. The existence of such ordered structures seems really to be correlated to the presence of trivalent iron. As we emphasized previously, the ionic size of Fe^{3+} and its isotropic electronic configuration ($3d^5$) make an octahedral or tetrahedral coordination equally probable. This is obviously a basic requirement of the model.

The investigations of other AMO_{3-y} systems without Fe^{3+} have been unsuccessful up to now. This model only seems to be applicable to certain Co^{3+} phases, due to the fact that strontium cobaltite $Sr_2Co_2O_5$ also exhibits a brownmillerite-type structure.

8 References

1. Bevan, D. J. M.: Non-stoichiometric compounds in Comprehensive Inorganic Chemistry. Pergamon Press 1973, p. 453
2. Le Roy-Eyring, O'Keeffe, M.: The Chemistry of Extended Defects in Non-Metallic Solids. North Holland, Amsterdam 1970
3. Schottky, W., Wagner, C.: Z. Phys. Chem. *B11*, 163 (1930)
4. Magnéli, A.: Arkiv. Kemi *1*, 513 (1950), Acta Cryst. *6*, 495 (1953)
5. Wadsley, A. D.: Rev. pure appl. Chem. *5*, 165 (1955)
6. Wadsley, A. D.: Non-Stoichiometric Compounds Mandelcorn (ed.). Academic Press 1964
7. Mrowec, S.: Rev. Int. Hautes Temp. et Réfractaires *14*, 4, 225 (1977)
8. Solid-State Chemistry, Proceedings of 5th Materials Research Symposium (N. B. S. Publication 364), 1972
9. Anderson, S., Wadsley, A. D.: Nature *211*, 581 (1966)
10. Spyridelis, P. Delavignette, P., Amelinckx, S.: Mat. Res. Bull. *2*, 615 (1967)
11. Bursill, L. A. et al.: Phil. Mag. *20*, 347 (1969)
12. Alpress, J. G., Gado, P.: Cryst. Lattice Defects *1*, 331 (1970)
13. Bursill, L. A, Hyde, B. G.: J. Sol. State Chem. *4*, 430 (1972)
14. Iijima, S.: J. Sol. State Chem. *14*, 52 (1975)
15. Alprèss, J. G.: J. Sol. State Chem. *1*, 28, 66 (1970); *2*, 78, 336 (1970)
16. Portier, R. et al.: Mat. Res. Bull. *9*, 371 (1974)
17. Chaminade, J. P., Pouchard, M.: Ann. Chim. *10*, 75 (1975)
18. Goodenough, J. B, Longo, J. M.: Crystallographic and Magnetic Properties of Perovskite and Perovskite-Related Compounds. Landolt-Börnstein, Neue Serie III 4a, Springer Verlag, Berlin, Heidelberg, New York 1970
19. Hagenmuller, P. (ed.): Preparative Methods in Solid-State Chemistry, Academic Press 1972
20. Negas, T., Roth, R. S.: J. Sol. State Chem. *1*, 409 (1970); *3*, 323 (1971)
21. Mac Carthy, G. J., White, W. B., Roy, R. J.: J. Amer. Chem. Soc. *52*, 463 (1969)
22. Kestigian, M., Dickinson, J. G., Ward, R.: J. Amer. Chem. Soc. *79*, 5598 (1957)
23. Shin, S., Yonemura, M., Ikawa, H.: Mat. Res. Bull. *13*, 1017 (1978)
24. Alario-Franco, M. A., Vallet Regi, M.: Nature *270* (5039), 706 (1977)
25. Tofield, B. C., Greaves, C., Fender, B. E. F.: Mat. Res. Bull. *10*, 737 (1975)
26. Jacobson, A. J., Horrox, J. W.: Acta Cryst. *B32* 1003 (1976)
27. Jabobson, A. J.: Acta Cryst. *B32*, 1087 (1976)
28. Goodenough, J. B.: J. Phys. Chem. Solids *6*, 287 (1958)
29. Takeda, T., Watanabe, H.: J. Phys. Soc. Japan *33*, 973 (1972)
30. Takeda, H., Shimada, M., Koizumi, M.: Mat. Res. Bull. *15*, 165 (1980)
31. Taguchi, H., Shimada, M., Koizumi, M.: J. Sol. State Chem. *29*, 221 (1979)
32. Banks, E., Berkooz, O., Nakagawa, T.: Solid-State Chemistry Proceedings of 5th Material Res. Symp., N. B. S. Publication 364 p. 265
33. Mac Chesney, J. B., Sherwood, R. C., Potter, J. F.: J. Chem. Phys. *43(6)*, 1907 (1965)
34. Zanne, M., Gleitzner, G.: J. Sol. State Chem. *6*, 163 (1973)
35. Browall, K. W., Muller, O., Doremus, R. H.: Mat. Res. Bull. *11*, 1475 (1976)
36. Van Buren, F. R.: J. Electroanal. Chem. *87*, 381 (1978)
37. Watanabe, H., Takeda, T.: Proc. of the Int. Conf. on Ferrites, Japan 1970 p. 588

38. Gallagher, P. K., Mac Chesney, J. B., Buchanan, D. N. E.: J. Chem. Phys. *41*, 8, 2429 (1964)
39. Gallagher, P. K., Mac Chesney, J. B., Buchanan, D. N. E.: J. Chem. Phys. *43(2)* 516 (1965)
40. Thomas, R., Clevenger, J. R.: J. Amer. Chem. Soc. *46*, 207 (1963)
41. Brixner, L. H.: Mat. Res. Bull. *3*, 299 (1968)
42. Oda, H. et al.: J. Phys. Soc. Japan, *42*, 101 (1977)
43. Yamamura, H., Kiriyama, R.: Bull. Chem. Soc. Japan *45*, 2702 (1972)
44. Tofield, B. C.: Reactivity of Solids. Wood, J. et al. (eds.) Plenum Press 1976, p. 201
45. Grenier, J. C. et al.: Mat. Res. Bull. *11*, 1219 (1976)
46. Grenier, J. C. et al.: J. Phys. *38*, 12 (1977)
47. Grenier, J. C. et al.: Mat. Res. Bull. *13*, 329 (1978)
48. Komornicki, S., Grenier, J. C., Fournes, L.: in press
49. Geller, S. Grant, R. W., Gonser, U.: Progr. Sol. State Chem. *5*, 5 (1971)
50. Nishihara, Y.: J. Phys. Soc. Japan *38*, 710 (1975)
51. Olazcuaga, R. et al.: J. Sol. State Chem. *13*, 275 (1975).
52. Wickoff, R. W. G.: Cryst. Struct. *3*, 101 (1965)
53. Grenier, J. C., Pouchard, M., Hagenmuller, P.: C. R. Acad. Sci. Paris *285 C*, 527 (1977)
54. Koehler, W. C., Wollan, E. O.: J. Phys. Chem. Solids *2*, 100 (1957)
55. Grenier, J. C., Pouchard, M., Hagenmuller, P.: Mat. Res. Bull. *11*, 721 (1976)
56. Bonnet, J. P. et al.: Mat. Res. Bull. *14*, 67 (1979)
57. Komornicki, S., Grenier, J. C., Pouchard, M., Hagenmuller, P.: Nouv. J. Chim. *5*, 161 (1981)

Received January 14, 1981
D. Reinen (editor)

Double-Double Effect of the Inner Transition Elements

Irena K. Fidelis and Tomasz J. Mioduski

Department of Nuclear Chemistry*, Institute of Nuclear Research, PL-03-195 WARSAW, Poland

Phenomenological, theoretical and practical aspects of the double-double effect are presented on the basis of the review of original papers concerning the variation with Z of thermodynamic functions of complex formation, lattice parameters, and several other properties of lanthanide and actinide compounds.

The importance of the analogy between the four segments: f^0–f^3, f^4–f^7, f^7–f^{10}, and f^{11}–f^{14}, besides double symmetry as the integral parts of the double-double effect is emphasized.

* Up to 1980 Department of Radiochemistry

1 Introduction

The double-double effect discovered in 1964 in this Department[1-3] is at present firmly established in the chemistry of f-elements[4-15]. Graphical presentation of the effect using the plot of a given property as a function of Z[1] resembles, to some extent, the crystal field effect observed within the d-element series, although each of these two effects is of different origin. The crystal field effect is usually presented in the form of two-humped plots of crystal lattice energy of enthalpy of hydration versus Z whereas the double-double effect in its full pattern, i.e. for the whole series of 15 f-elements (La-Lu, or Ac-Lw), is composed of two main branches joined at the f^7 configuration, each of which consisting of two segments: f^0–f^3, f^4–f^7, and f^7–f^{10}, f^{11}–f^{14}, respectively. In other words, the main division is done by f^7 configuration followed by subdivision of subgroups f^0–f^7, and f^7–f^{14} by the central pairs f^3–f^4 and f^{10}–f^{11}, respectively. However, a plot of differences in a given property observed along the series of f-elements is a more precise representation of the double-double effect. In this form, the effect has been discovered[1)] owing to the method of reversed phase-partition chromatography[2] which enabled the determination of accurate values of consecutive separation factors. The separation factor, β, defined as the ratio of extraction coefficients of two neighboring lanthanides, has been used as a classical measure of differences in extractability or complex formation ability of consecutive f-elements. The plot of β values versus consecutive pairs of neighbouring lanthanides shows the periodic sequence of four maxima and four minima which, because of the known relation

$$\Delta G^0_{Z+1} - \Delta G^0_Z = -RT \ln \beta \ , \qquad (1)$$

implies a periodicity of a ΔG^0 vs Z plot consisting therefore of four branches. The plot of differences in free energy changes determined for consecutive pairs of neighboring lanthanides, $\Delta G^0_{Z+1} - \Delta G^0_Z = \Delta(\Delta G^0)$, for HEHØP as a ligand and the plot of individual values of relative changes in free energy versus Z, are presented in Fig. 1. The upper plot is important because it defines the shape of the segments: the decrease of differences observed for neighboring elements within each of four segments (La–Nd, Pm–Gd, Gd–Ho, and Er–Lu) is thus revealed. The lower plot taken from the second paper[16] devoted to the problem is that one which Peppard et al.[4,5] have used for characterizing the phenomenon as "tetrad effect".

The double-double effect has been considered in review articles[17-20] lectures[21-24] and also in some monographs [10,25-27]. In these publications, however, emphasis is placed on the double-double effect observed within the lanthanide series. The purpose of this paper is to review the literature data on the double-double effect observed in actinides. In contrast to lanthanides, in the case of actinides, the double-double effect cannot be observed in its full pattern because of the following reasons: (i) the elements heavier than einsteinium are still hardly available, (ii) the stable oxidation state changes along the series, and (iii) there are only few data on stability constants or extraction coefficients for

1 Instead of a more precise plot, i.e. the plot of differences in this property as a function of consecutive pairs of Z[1)]

2 Recently known as extraction chromatography

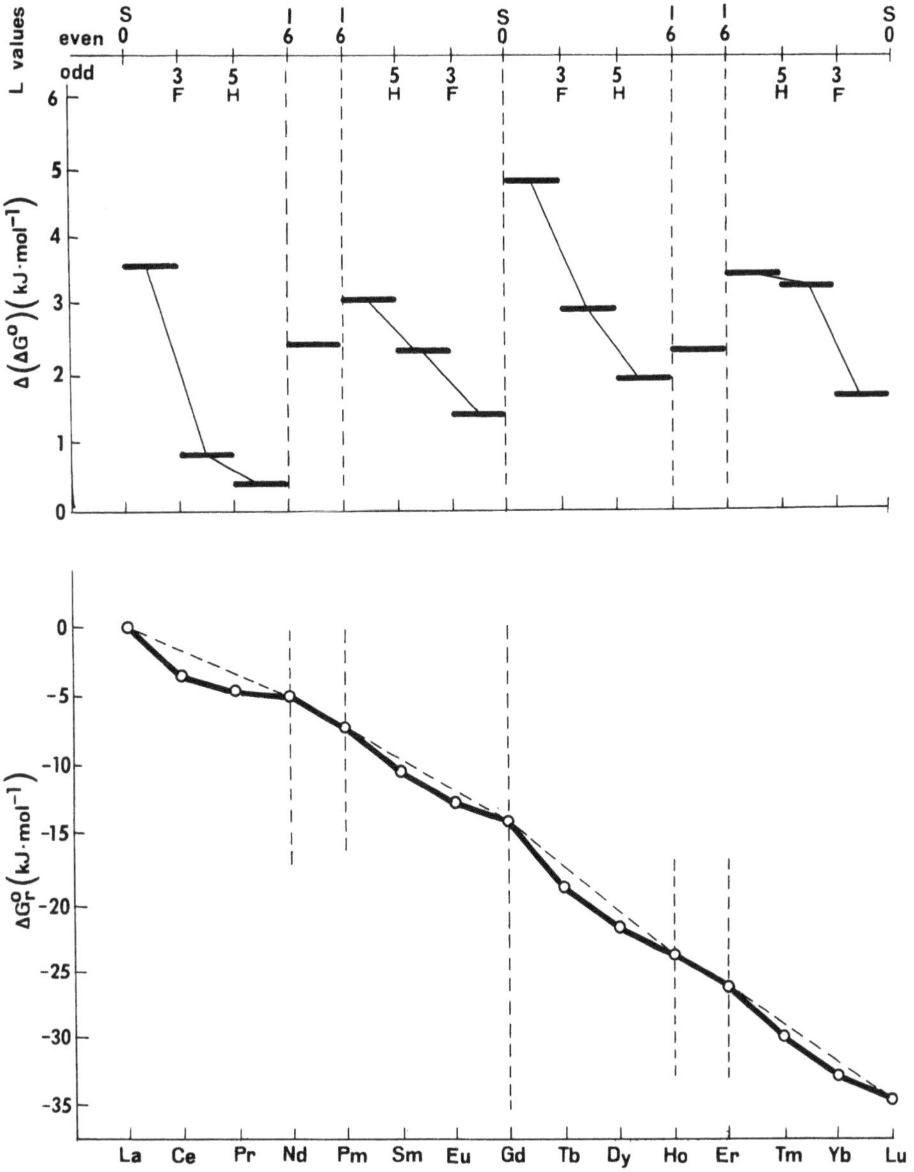

Fig. 1. Double-double effect described by the differences in the ΔG^0 values for consecutive pairs of neighboring 4f-elements (upper plot) and by the ΔG^0 vs Z plot (lower plot). Experimental values of free energy changes for HEHØP as a ligand have been used[16]

actinides in oxidation states higher than + 3. At the same time, however actinides offer a good possibility of studying the double-double effect for various valence states. First such an example has been presented for lattice parameters of isostructural sequence of the monoclinic actinide tetrafluorides[2]. Using the data listed on the first positions of the second column in Table 1, in the paper of Kennan and Asprey[28], the "a_0" parameter for

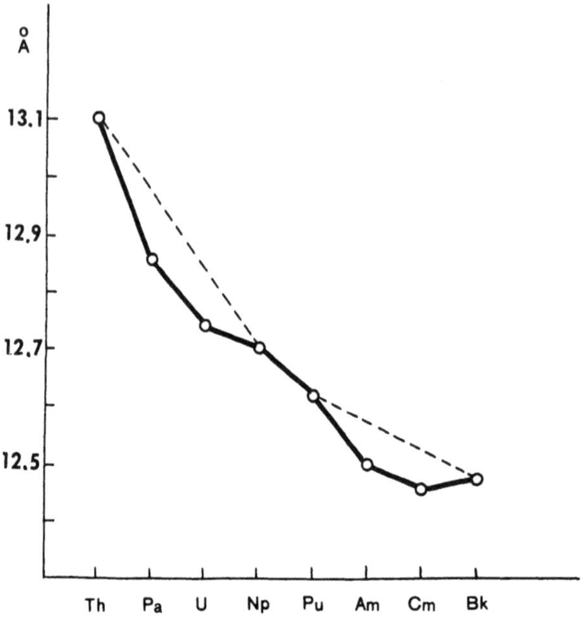

Fig. 2. Double-double effect observed in parameter "a_0" determined for the isostructural sequence of monoclinic actinide tetrafluorides: data for the Th–Bk (f^0–f^7) range, i.e. for one half of the full pattern of the effect

eight tetrafluorides of consecutive elements, from Th(IV) to Bk(IV), has been plotted as a function of Z. Most of the used a_o values are taken from the papers of Asprey et al. or Zachariasen. The values of interest include: ThF_4: 13.1[29], PaF_4: 12.86[30], UF_4: 12.73[31, 32], NpF_4: 12.70[29], PuF_4: 12.62[29], AmF_4: 12.49[33], CmF_4: 12.45[34], BkF_4: 12.47[35]. The plot is presented in Fig. 2. It shows exactly one half of the full pattern of the effect. For the tetravalent state the positions of segments are shifted by one place in comparison with those for the trivalent state. The points of Pa, U, Am and Cm are distinctly shifted below the interpolated dashed lines. The mean distance is equal to about 20% to the total decrease observed from Th to Bk. This could suggest that the effect is greater in the case of the 5 f series than in the 4 f series. Thus, the comparison of the 5 f with the 4 f series is of interest both from phenomenological and theoretical points of view.

2 The Nature of the Double-Double Effect

That the double-double effect originates in the various f-electron configurations was stated on the basis of the correlation existing between the full pattern of the effect and the sequence of values of the L-quantum number[2]. The correlation consists in the occurrence of the same double symmetry in both 1) the series of L-quantum number values of the ground terms of the f-element ions and 2) the sequence of relatively stabilized or destabilized f-electron configurations, i.e. the double-double effect. We assumed the f^0, f^3, f^4, f^7, f^{10}, f^{11}, and f^{14} configurations as relatively stable whereas the four pairs f^1–f^2, f^5–f^6, f^8–f^9 and f^{12}–f^{13} are assumed to be relatively destabilized configurations.

Relatively more stable configurations have even values of L-quantum number (S = 0 and I = 6 terms) while relatively unstable configurations correspond to odd values of the L-quantum number (F = 3 and H = 5 terms). It should be noticed that relatively stabilized configurations display a relatively smaller ability to complex formation (less negative values of ΔG^0) and vice versa. In Fig. 1 it is shown that the sequence of L-values has the same double symmetry as the double-double effect. However, no linear correlation has been experimentally observed between changes in the free energy of complex formation or lattice parameters of f-element compounds and the values of the L-quantum number of appropriate f-ions. The lack of such linear correlation is explained in Sect. 5.

That the double-double effect is an intrinsic feature of f-electron configurations is also confirmed by the fact that the pattern of the effect observed in the case of tetravalent actinides starts with Th(IV), i.e. with the $5f^0$ configuration, and is consequently shifted by one place. For hexavalent actinides, the double-double effect most probably starts with uranium ($5f^0$) encompassing the U(VI)–Am(VI) tetrad for the first segment (see Sect. 4).

An excellent confirmation that the double-double effect is an intrinsic feature of f-orbitals has been reported by Jørgensen[36] who has shown the reasons for the singularities of the 3–4, 7 and 10–11 f-electron configurations on the basis of the theory of spin-pairing energy[37–40]. This theory was based on the observation that for a given configuration l^q, the energy difference between the center of gravity of all (S–1) terms and the center of gravity of all S terms is equal to 2 DS where D is a linear combination of interelectronic integrals in the Slater and Condon – Shortley theory. The lowest term having the maximum value of S is situated below the baricenter of the whole configuration to the extent of

$$-4(q^2-q)\,D/13 \qquad (2)$$

for $q \leq 7$, while for the other configurations, q is replaced by 14–q. Expression (2) is a spin-pairing energy where D denotes a spin-pairing energy parameter[37] which is assumed to be equal to about 6500 cm^{-1} for 4f the series and about 3900 cm^{-1} for the 5f group[36]. For configurations f^q for which more than one term represents the maximum of S the lowest term is further stabilized:

$$\begin{array}{ll} -21\,E^3, \text{ for} & q = 3,\ 4,\ 10 \text{ and } 11 \\ -9\,E^3, \text{ for} & q = 2,\ 5,\ 9 \text{ and } 12 \end{array} \qquad (3)$$

where Racah's parameter E^3 is about a tenth of D.

Thus, the stabilization of the ground electronic states of a given f^q configuration relative to the baricenter of the whole configuration is as follows:

configuration	term	stabilization energy
f^2, f^{12}	^3H	$-8\,D/13 - 9\,E^3$
f^3, f^{11}	^4I	$-24\,D/13 - 21\,E^3$
f^4, f^{10}	^5I	$-48\,D/13 - 21\,E^3$
f^5, f^9	^6H	$-80\,D/13 - 9\,E^3$
f^6, f^8	^7F	$-120\,D/13$
f^7	^8S	$-168\,D/13$

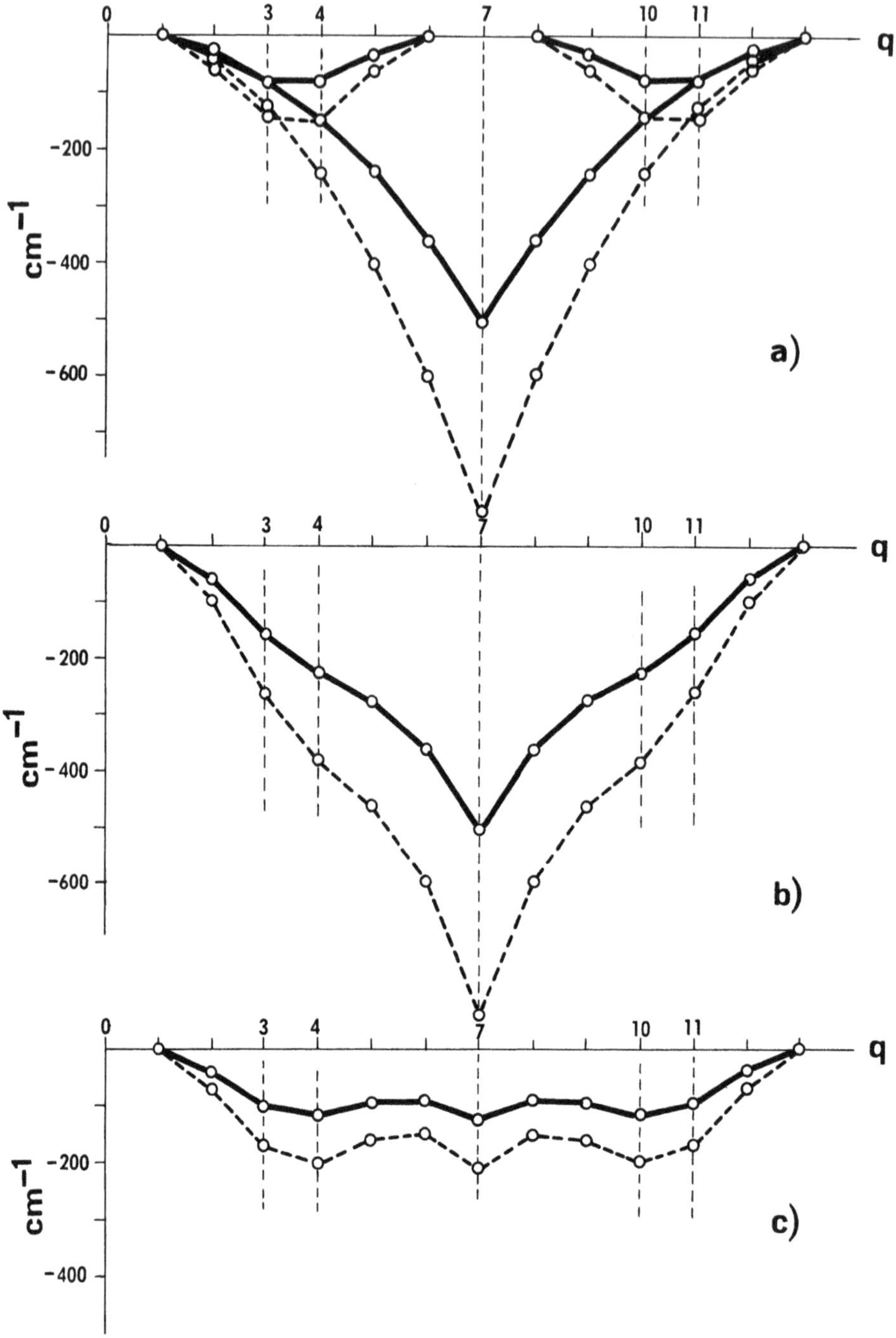

Fig. 3 a–c. Double symmetry displayed by the spin-pairing stabilization energy: **a)** Symmetry with respect to the f^7 configuration according to expression (2), the symmetry within the two subgroups according to expression (3). The values plotted are calculated assuming that both parameters D and E^3 decrease by one percent; **b)** combined stabilization; **c)** combined stabilization under the assumption that parameter E^3 decreases by one percent while D only by 0.25 percent. (———— 5 f, ------ 4 f)

No stabilization of this kind occurs for configurations f^0, f^{14} (^1S terms) as well as for f^1 and f^{13} (^2F terms).

In his note[36], Jørgensen attempted to give a theoretical explanation of the double-double effect which is assumed to result from the decrease of phenomenological parameters of interelectronic repulsion with increasing covalency of the metal-ligand bonding. There is a close relation between the extent of covalent bonding and the nephelauxetic effect which becomes the more pronounced, the higher the reducing power of the ligands and the higher the oxidizing power of the central ion[37]. The electron cloud expansion of the central ion under the influence of ligand electrons, known as the nephelauxetic effect, has been the subject of many of Jørgensen's papers[41–49].

The coefficients of D-dependent terms display symmetry with respect to the f^7 configuration, and simultaneously division of the whole 4f- or 5f-series into two the subgroups f^0–f^7, f^7–f^{14}. Changes in coefficients of E^3 show the symmetry within the two subgroups. This is described in Fig. 3a, assuming that both parameters of interelectronic repulsion, D and E^3, decrease by one percent. The combined stabilization is plotted in Fig. 3b. Superposition of stabilization originating from expression (3) in the plot of expression (2) as a function of q gives the singularities of f^3–f^4 and f^{10}–f^{11} configurations, i.e. two central pairs within the subgroups f^0–f^7 and f^7–f^{14}. The central position of the f^7 configuration is obviously still valid. Furthermore, since the differences between the complexes under investigation and aquoions are generally measured experimentally, there are reasons for assuming that E^3 shows a more pronounced nephelauxetic effect on the ground state energy than D[36]. If E^3 is decreased by one percent while D only by 0.25 percent, the combination of expression (2) and (3) gives the plots presented in Fig. 3c. Singularities of the central pairs f^3–f^4 and f^{10}–f^{11} are now more pronounced in both series of f-elements. However, it should be noted that in all three cases described in Fig. 3, plots for 4f series are more pronounced than those for 5f series because the numerical values of parameters D and E^3 are greater for lanthanides(III) than for actinides(III). In summary, the stabilization of the ground terms relative to the baricenter of the whole configuration, plotted here as a function of q, shows the same double symmetry as the double-double effect. This double symmetry can be observed in the sequence of the ground state terms as regards the values of the total orbital angular momentum, L, of both series of f-elements[2]. It should be noted that the numerical values of L, plotted as a function of q, show two maxima[10]. These are two central pairs being responsible for the symmetry within the two subgroups f^0–f^7 and f^7–f^{14}. Obviously, the symmetry with respect to the f^7 configuration also occurs in this plot, namely around two central pairs, within two subgroups, the symmetrical values of H and F terms decreasing to the value equal to zero (S terms) for configurations f^0, f^7 and f^{14}. The S term of the f^7 configuration is a common point for both subgroups. On the other hand, the variation of the S quantum number plotted as a function of q, represents a maximum value for $q = 7$, and a symmetry relative to the f^7 configuration[10]. Both the L vs q and S vs q plots are illustrated in Fig. 4.

Returning to coefficients of interelectronic repulsion parameters D or $E^1 = 8D/9$ and E^3, the explanation of double symmetry reflected in the double-double effect has been advanced also by Nugent[50].

The problem of the comparison of the magnitude of the double-double effect in the case of (a) lanthanide(III) and actinide(III) ions as well as in the case of (b) actinides in + 3 and higher valence states can be discussed on the basis of the following analysis.

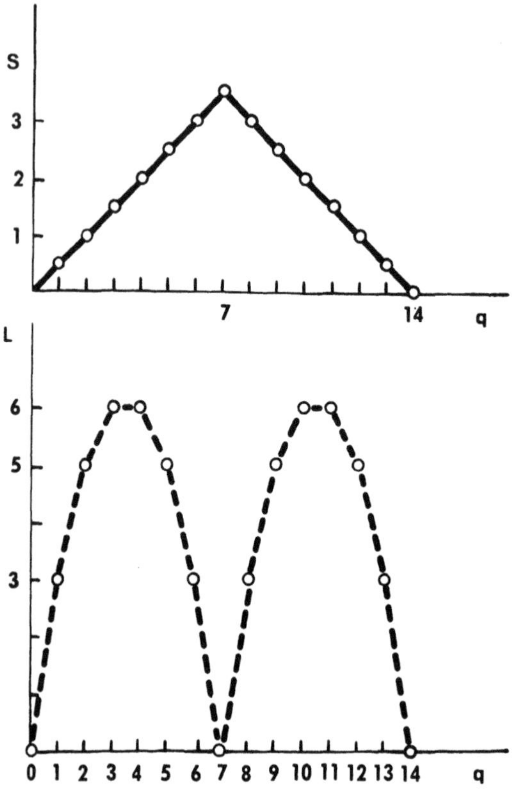

Fig. 4. Numerical values of quantum numbers L and S plotted as a function of q

Let us take into account the double-double effect in the case of free energy of complex formation. Then, the experimentally measured effect can be considered as a difference between the stabilization energy of the ground-state term for the complex and that for the aquoion. The experimentally observed double-double effect is the greater, the more pronounced the change of parameters E^1 and E^3 under the nephelauxetic action of complexing ligands. Thus, the influence of the nephelauxetic action of ligands expanding the partly filled shell of the central ion and, at the same time, decreasing the parameters of interelectronic repulsion E^1 and E^3, is of basic importance to both comparisons (a) and (b). Some ligands can produce a comparatively strong decrease of interelectronic repulsion parameters of 5 f-ions whch can result in a stronger double-double effect in the 5f than in the 4f series because of a greater expansion of the 5f electron cloud. Thus, although the numerical values of parameters D and E[36] are larger for lanthanide(III) than those for actinide(III) ions, one can observe greater double-double effect for the latter series than for the former one. Similar factors can cause a stronger double-double effect in the case of actinide(III) complexes than in the case of complexes of actinides in higher oxidation states, although parameters D and E^3 of actinide(III) ions are smaller than those of actinides in higher oxidation states.

3 The Double-Double Effect in Complex Formation and Extraction of Trivalent Actinides

Siekierski has given a survey[51)] of the stability constants β_3 of actinide(III) complexes in the Am-Md range with 8 ligands, i.e. citrates, lactates, glycolates, α-hydroxyisobutyrates, 2-ethylhexyl phenylphosphonates, di(2-ethylhexyl) phosphates, dibutyl phosphates, and 2-thenoyl trifluoroacetonates. He calculated the mean separation factor[3], $\bar{\alpha}$, for each pair of adjacent actinide(III) ions, according to the definition:

$$\bar{\alpha} = \frac{\sum_{1}^{n} = \frac{(\beta_3)_{Z+1}}{(\beta_3)_Z}}{n} \tag{4}$$

where Z and n denote the atomic number and the number of ligands, respectively. This kind of a statistical approach has first been made for lanthanides[51)]. The results obtained for actinide(III) compounds together with those for the respective range of lanthanide(III) compounds are presented in Fig. 5. The upper plot illustrates the pattern of the double-double effect (in the range of f^6–f^{14}) expressed by statistical data on lanth-

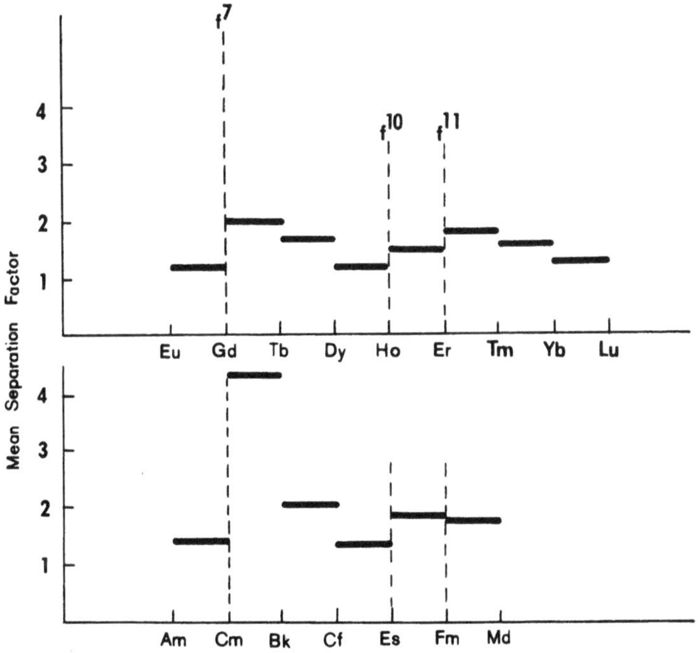

Fig. 5. Double-double effect observed in the mean separation factor for 4f (upper plot) and 5f (lower plot) trivalent elements

3 In the paper of Siekierski[51)] the mean separation factor is denoted by $\bar{\alpha}$, to avoid confusion with β_3 stability constants

anides for over 30 ligands. The lower plot concerning actinides shows the pattern of the double-double effect from the beginning of investigated range up to fermium. The value obtained for the last investigated pair, Fm–Md, is too small to fit the pattern of the double-double effect. It should be mentioned however, that the mean separation factor for this pair has been calculated from five sets of data only whereas for the majority of investigated pairs as much as fifteen sets of data have been used for the calculation. A comparison between the plot for actinides and that for lanthanides reveals the same sequene of regular variations in the mean separation factor (double-double effect) for both series of analogous f-elements, i.e. the elements with the same number of f-electrons, except for the Fm–Md pair for which the result has a rather poor statistics. On the other hand, the plots in Fig. 5 show a more developed third segment (Cm–Es) for 5f series than that for 4f series (Gd–Ho). From this comparison, however, one cannot infer that the effect is generally larger for actinides than for lanthanides. From the very beginning[1)], it was pointed out that for such ligands as HEHØP, HDEHP and EDTA, the effect is much more pronounced than for such ligands as TBP, NTA or HIB. Therefore, while attempting to express the magnitude of the effect it was assumed that there are two classes of ligands, the first class for which a comparatively large double-double effect is observed and the second class for which less fluctuations in the separation factor or less developed segments occur [10)] [4]. The most probable reason why statistical results presented in Fig. 5 show a weaker double-double effect for the 4f than for the 5f series is due to the fact that for lanthanides the data for the second class ligands were mainly taken for calculations while for actinides the first class ligands predominated, especially in the Cm–Cf range.

Statistical results obtained for the 5f series are constantly supplemented by new experimental data, e.g. by the data of Harmon et al.[52)], who studied the formation of isothiocyanato complexes of trivalent Am, Cm, Bk, Cf, and Es. They applied a solvent extraction technique using HDEHP as an extractant in the medium NaSCN–$NaClO_4$. The double-double effect observed in the variation of the β_3 stability constants with Z has been explained by changes, occurring during complexation in the inner coordination sphere of actinide(III) ions. On this basis it was tentatively concluded that the neutral $An(NCS)_3$ complexes are of the inner sphere type. The magnitude of the effect[5] whch could be estimated to be equal to about −350 and −200 cal/mol for Bk and Cf respectively, shows a significant decrease of interelectronic repulsion parameters as a result of the nephelauxetic action of ligands, accompanying the process of extraction. Guillaumont et al.[53)] have shown the existence of the third tetrad (Cm^{3+}–Es^{3+}) of the double-double effect in the variation with Z, of the first hydrolysis and extraction constants, for the complexation of actinide(III) ions with TTA. Guillaumont et al. have also detected the double-double effect in the Am–Es range, in the variation of the β_2 stability constant for citrate complexes of trivalent actinides[54)]. On the basis of the cocrystallization of the ethyl sulfates of all lanthanides and some actinide(III) ions Mioduski and Siekierski have predicted the possibility of the occurrence of the double-double effect in the enthalpy and free energy of hydration of the actinide(III) series[55, 56)]. Because the coordination

4 Ref. 10, Sect. 2.5, p. 18, Figs. 2.1 and 2.2

5 Measured as the distance between experimental points for destabilized configurations (f^8 of Bk(III) and f^9 of Cf(III) within the segment of interest) and the interpolation line connecting the f^7–f^{10} points

number of actinide(III) aquoions decreases from 9 to 8, presumably somewhere about Fm, and because this decrease purports a shift of the position of ligands in the nephelauxetic series, it was predicted [57, 58] that the double-double effect in thermodynamic functions of hydration of actinide ions is much more pronounced for the last segment (Fm–Lw) than for the previous ones.

In the case of aquoions of lanthanides the coordination number decreases before Gd(III). Thus, the double-double effect observed in the hydration data is more pronounced for the Gd–Ho and Er–Lu segments than for the La–Gd subgroup. This conclusion has also been drawn by Guillaumont et al.[13, 59].

4 The Double-Double Effect in Lattice Parameters and Unit Cell Volumes of Actinides

Probably the best example of the double-double effect in actinides is that resulting from changes in lattice parameters of monoclinic actinide tetrafluorides described in Fig. 2. The most interesting feature in this case is the fact that the effect is so distinct although the fluoride ligand is a very "hard" Pearson base nearly starting the nephelauxetic series[37]. One could assume that for strictly ionic tetrafluorides some relativistic effects are involved in the double-double effect, due to the large atomic numbers and high concentration of the +4 cations. Pyykkö[60] has presented the problem of the lanthanide contraction as a relativistic effect. He performed relativistic Hartree-Fock (Dirac-Fock) one-center expansion calculations on the tetrahedral-covalent model hydrides CeH_4 and HfH_4, using these hydrides as model systems for the covalent bonds of the central metal atom. The theoretical bond lengths were obtained from the Morse potentials. Their difference gives a lanthanide contraction which is in good agreement with experiment. The conclusion drawn is that the nonrelativistic lanthanide contraction is 86% of the total one. So, the lanthanide contraction seems to be mainly, although not entirely, a nonrelativistic shell-structure effect. This conclusion does not deny our opinion that the double-double effect observed in lattice parameters is unusually pronounced because in this case, the changes determined for consecutive elements are measured directly and not as differences between the values for the complex and respective aquoion, as is the case of complex formation and extraction data.

A systematic study of the mean specific unit cell volumes of actinides in the III, IV and VI oxidation states has been done by Siekierski and one of the authors[61, 62] using the same statistical approach as previously described for stability constants[51]. The results are illustrated in Fig. 6. The plots represent the variations of the mean specific unit cell volume, ν, as a function of the number of f-electrons[6] for lanthanides(III) and actinides in various valence states. Curve 1 is shown for comparison. This curve has been plotted on the basis of as many as about 60 different lanthanide(III) compounds[63]. It shows the full pattern of the double-double effect represneting two branches joined at the central f^7 configuration, each of them consisting of two segments connected by the central pairs f^3–f^4 and f^{10}–f^{11}, respectively (see lower plot in Fig. 1). Curve 2 is a summary curve for all

6 Supposed to be present in the consecutive actinide or lanthanide ion

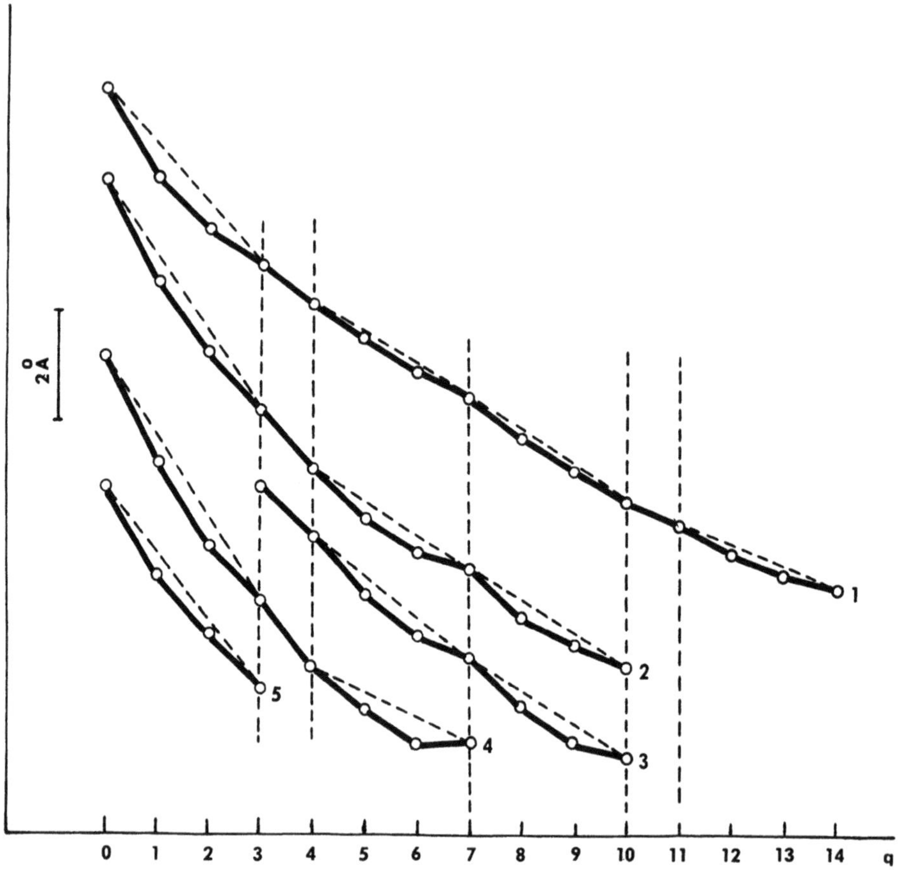

Fig. 6. Variation of the mean specific unit cell volume with the number of f-electrons. Curve 1 – data for lanthanides(III); curve 2 – summary curve for actinides; curves 3–5 – III-, IV- and VI-valent actinides, respectively

data for actinides. The mean specific unit cell volume, i.e. the volume per actinide ion, ν, used for plotting curves 3–5, has been obtained by dividing the unit cell volume by the number of actinide ions contained in it. Curve 3 showing the results for trivalent actinides has been plotted using the experimental data on 8 different compounds: oxides, sulfides, fluorides, chlorides, hexahydrates of bromides, oxychlorides, phosphates and niobates. Curve 4 illustrating the results for tetravalent actinides is based on experimental data for such compounds as oxides, sulfoxides, chlorides, fluorides, and complex fluorides ($LiAnF_5$, Rb_2AnF_6, $Na_7An_6F_{31}$, $K_7An_6F_{31}$, and $Rb_7An_6F_{31}$). Curve 5 showing the results for hexavalent actinides is based on the data obtained for Li_4AnO_5, Na_4AnO_5, AnO_2F_2, and $NaAnO_2(Ac)_3$. For this oxidation state, the double-double effect starts with uranium ($5f^0$ configuration) and is limited to the range of four atomic numbers only because of difficulties in the availability of the hexavalent state in actinides heavier than americium. Although the data for each oxidation state are limited, it can be easily seen that the changes in ν as a function of the number of 5f electrons exhibit the same pattern independently of the oxidation state. Because of this, and because of the fact that the values of ν for different oxidation states are comparable, the mean specific unit cell volumes for all investigated valence states have been calculated to obtain a summary curve extended to the full available range of the 5f series. As it can be seen in Fig. 6, the summary curve is similarly to that describing the changes for lanthanides. Although data for the last segment are lacking, curve 2 represents the double-double effect for the 5f seires which, similar to the 4f series, can be divided into two subgroups having in common the $5f^7$ configuration. The first subgroup is further subdivided into two segments connected by the central pair f^3–f^4. The second subgroup only contains data for the first segment f^7–f^{10}.

From the results shown in Fig. 6, some conclusions may be drawn. A comparison of the lanthanide curve with the summary curve for actinides suggests that the double-double effect is more pronounced in actinides than in lanthanides. A comparison of curves 3 and 4 reveals that the effect is the stronger, the higher the oxidation state of the actinide ions. Moreover, curves relating to actinides show, that the mean specific unit cell volumes decreases with Z more rapidly for this series than for lanthanides. This suggests that the actinide contraction is larger than that of lanthanides. Such a conclusion also results from the data of Zachariasen[64].

5 Some General Remarks

The schematic pattern of the double-double effect is presented in Fig. 7. In this figure the two halves of the effect are superimposed, i.e. the f^0–f^7 and f^7–f^{14} subgroups are described by one plot, as it was done in the first paper on the effect in actinides[61]. The upper part of Fig. 7 shows the changes in a given f-electron element property, A, depending on the course of the series. The lower part of Fig. 7 illustrates variations in the differences between property A for neighboring elements observed along the series. This is a schematic pattern of the original representation of the effect which is of importance because it emphasizes the analogy between two subgroups as well as between four segments revealing the decrease of the differences within the f^0–f^3, f^4–f^7, f^7–f^{10}, and f^{11}–f^{14} ranges. At the same time, it emphasizes the regularities in the changes of the properties of f-elements, consisting in the periodic decrease of consecutive differences of a given property. The analogy occurring between two subgroups as well as between four segments of the f-element series, in the form of A vs Z plots, is schematically described in Fig. 8. If property A decreases with Z, or with the number of f-electrons, as is true for most experimentally observed cases, then the curvatures of segments are directed downward. If property A increases with Z, the curvatures of segments are directed upward, as it results from a plot of stability constants or relative extraction coefficients against Z[10,65]. As it can be seen from Fig. 8, the double-double effect consists in the relative stabilization of the f^3, f^4 – and f^{10}, f^{11} – electron configurations, in addition to the earlier known stabilization of configurations f^0, f^7 and f^{14}. The existence of distinguished configurations results in 1) the main division of both f-element series into two subgroups of analogous elements, f^0–f^7 and f^7–f^{14}, and 2) subdivision of these subgroups into four segments by the central pairs f^3–f^4 and f^{10}–f^{11}. From a phenomenological point of view, it

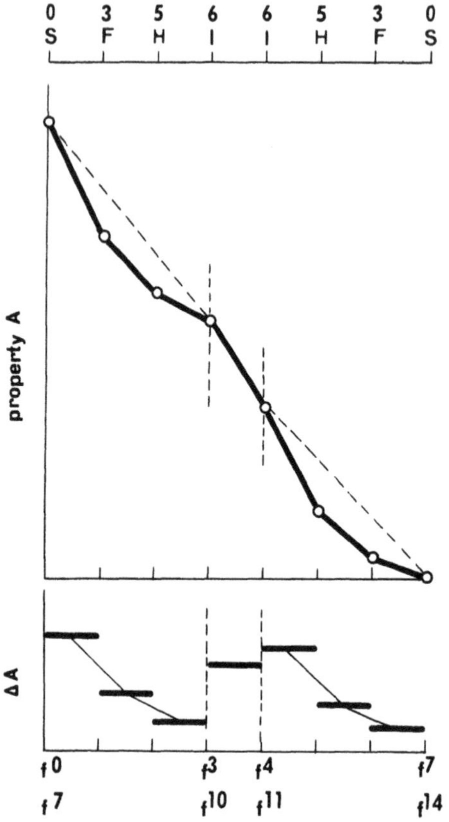

Fig. 7. Schematic pattern of the double-double effect as a plot of property A vs q (upper part) and of the differences in property A vs consecutive pairs of f-elements (lower part)

should be noted that in most cases the main classification into two subgroups is more distinct than the subdivision into four segments[51].

Double symmetry exhibited in the location of distinguished configurations as well as symmetrical but not equivalent double division of the series, prompted us to introduce the term "double-double effect"[2,3]. The introduction of this term is mainly due to the fact that the double symmetry reflected in the properties of f-electron elements is the same as that exhibited by the sequence of their free-ion ground terms as regards the values of the L-quantum number. At the same time, it should be emphasized that the main significance of the correlation, existing between the sequence of L-values and the sequence of the relative stability of f-electron configurations, consists in the fact that it is the first and strong evidence that the double-double effect is an intrinsic feature of f-electron configurations[2]. It should, however, be mentioned that unfortunately some misleading hypotheses regarding the correlation between the properties of f-elements and the L-quantum numbers have been offered or suggested[66,67].

Double symmetry existing in the sequence of L-values and reflected in the double-double effect, i.e. in many experimentally observed properties of f-elements, known as the correlation between the f-element properties and the sequence of the ground state L terms[2], constitutes a problem which should not be underestimated. Besides double symmetry, the analogy of changes in a given property existing between two subgroups as

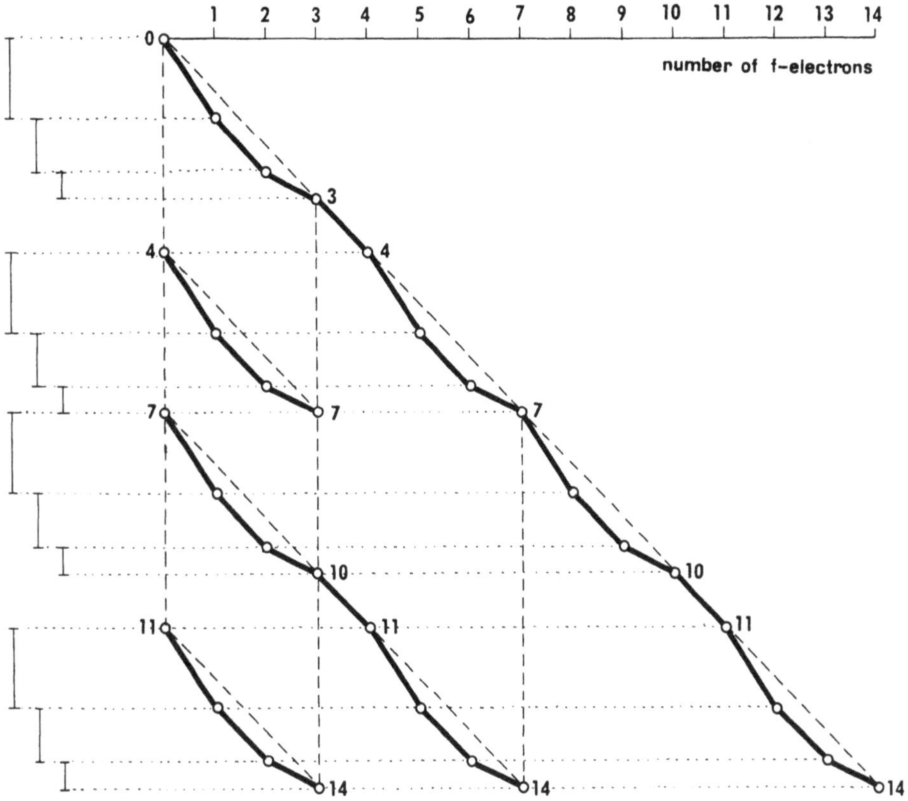

Fig. 8. Scheme of the analogy observed between the two subgroups f^0–f^7 and f^7–f^{14} as well as between the four segments f^0–f^3, f^4–f^7, f^7–f^{10}, and f^{11}–f^{14}

well as between four segments is a feature of f-elements. The analogy of changes in a given property means the analogy between differences observed for neighboring elements going across the series. As it is depicted in Fig. 8 the decrease of differences in a given property of consecutive elements, repeated by each of the four segments, is a characteristic feature of the f-element series. Thus, no symmetry can generally be observed when the sequence of differences in property A of neighboring elements is concerned. Such a symmetry with respect to the f^7 configuration as well as to the central pairs f^3–f^4 and f^{10}–f^{11} of both subgroups f^0–f^7 and f^7–f^{14} is absolutely required for the validity of the "inclined W" hypothesis. The idea of "inclined W" plots has been reported as follows: "The properties of the lanthanide and actinide ions and their complexes vary linearly with the L-values (the total angular monumentum). Usually, plots resembling a four-segmented "inclined W" are obtained for the whole series, where the data are available. Although some deviation from a symmetrically "inclined W" plot occurred in a few cases, the linearity is maintained within the four segments"[66]. In the same article, it has been suggested that "for correlating the properties of the lanthanide and actinide series, the L-values rather than the atomic number Z, or number of the f-electrons, should be used".

An attempt to elucidate the problem has been made in a short note[68], pointing out that the "inclined W" hypothesis is at variance with the double-double effect based on ample experimental material[1, 10, 51, 61, 63, 65]. In this note three sets of experimental data has been selected or chosen to illustrate how far from "linearity" the experimental points, obtained from typical systems studied, are. Two of these sets include thermodynamic data for lanthanides on the extraction by HEHØP[69], the extractant belonging to the first class of ligands, and TBP[6] representing the second class of ligand (see Sect. 3). As the third set, data for tetravalent actinides have been chosen. These are data used for plotting the pattern of the double-double effect presented in Fig. 2 (see Sect. 1). In this case half of the series is demonstrated. All three sets of data have been plotted as a function of L-values to try to fit the linearity. This is shown in Figs. 9–11. The consecutive values of the total angular momentum are as follows: 0, 3, 5, and 6, for the first segment consisting of four elements, then 6, 5, 3 and 0 for the second segment. Starting with the central value of 0 (S-term of the f^7-configuration), the same sequence is repeated for the next two segments. Thus, a proportionality of experimental data with respect to the differences between consecutive L-values, equal to 3, 2, 1 and 1, 2, 3 for the first and second segment, respectively (and the same for the second, f^7–f^{14}, subgroup), is necessar-

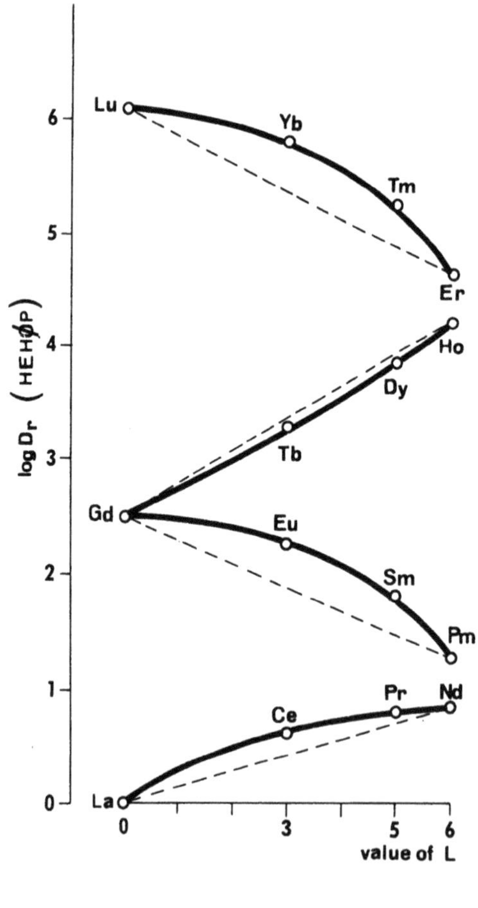

Fig. 9. Plot of the logarithm of the relative extraction coefficient vs L-quantum number for the extraction of trivalent lanthanides by HEHØP. Experimental data of static extraction at 25 °C[69]

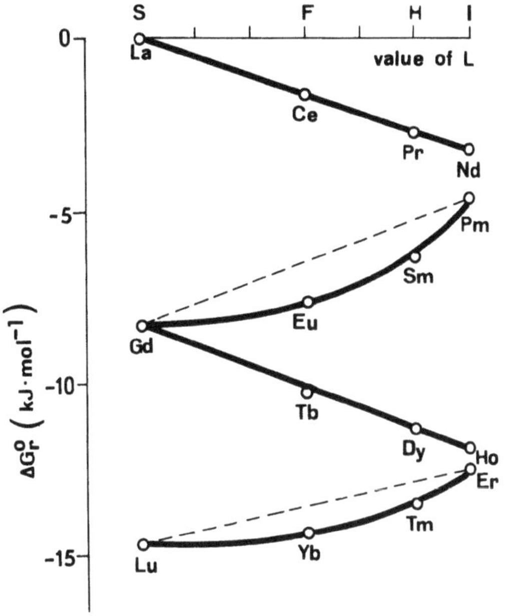

Fig. 10. Plot of the relative free energy changes vs L-quantum number for the extraction of trivalent lanthanides by TBP. Data obtained at 25 °C[6])

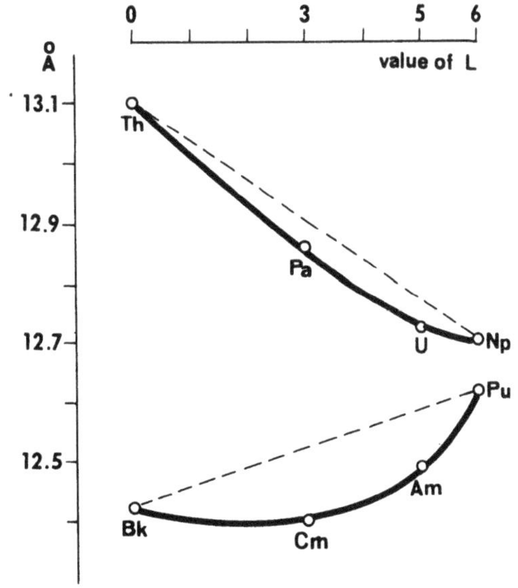

Fig. 11. Plot of lattice parameter "a_0" vs L-quantum number for isostructural monoclinic actinide tetrafluorides. Data reported by Keenan and Asprey[28])

ily required by the hypothesis of the "linear correlation of the properties of f-transition ions with their L-quantum number"[67]). In other words, the differences between the consecutive experimental values should decrease within the first segment, then increase up to the central f^7 configuration, and again the same pattern is required for the realization of the second subgroup of the series. However, the experimental data generally show, somewhat different patterns. If the schematic pattern of changes in the properties of f-elements was represented as a function of L-quantum numbers, then this pattern would have the shape indicated by the open circles connected with the solid lines in Fig. 12. This is the pattern of periodically decreasing differences plotted as a function of L-values. Two conditions are of importance from this point of view. The first condition is that within the second and fourth segment, experimental points cannot follow the straight lines if they are not at variance with the double-double effect and also with the earlier known half-filled shell effect[70]. The second condition is that within the first and third segment, experimental points may, although not necessarily, follow the straight lines. It should however be mentioned, that the only reason for linearity, which may occur within the first and third segment, is the coincidence of the L-scale with the decrease observed within the four segments. It should also be noted that the half-filled shell effect known since at least 1939[71] has been confirmed by numerous experimental data[72–79] before the double-double effect has been discovered. After that, experimental data illustrating the double-double effect at the same time confirm the half-filled shell effect which is included in the former one. Returning to experimental data chosen for the illustration of the changes in the f-element properties as a function of L-values, one should mention that

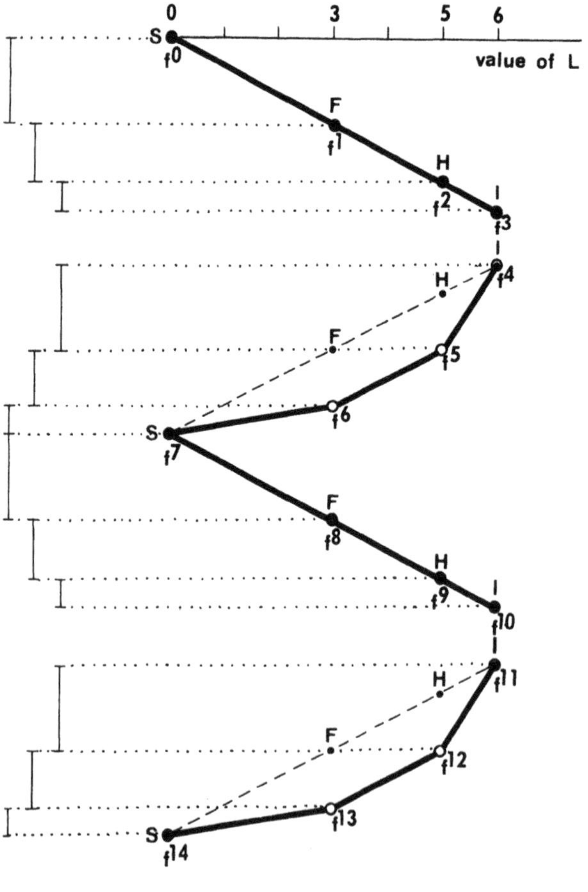

Fig. 12. Schematic pattern of changes in the f-element properties plotted as a function of L-values (open circles connected with the solid lines). Dashed lines connecting full points within the second and fourth segments show the "inclined W" plot

the data on lanthanides relate to classical ligands in this field. The ligands include HEHØP, for which the double-double effect has been discovered[1)], and TBP, the typical representative of the second class ligands (see Sect. 3). A similarity shown by plots in Figs. 9 and 10, with respect to the schematic pattern in Fig. 12, indicates 1) that the pattern is a good approximation to the real picture, and 2) that experimental data reveal no linearity within the f^4–f^7 and f^{11}–f^{14} segments. Such a similarity can also be observed in the case of the plot given in Fig. 11, as well as in the case of any series of experimental data representing the double-double effect. Sources of such experimental data are papers dealing with the problem of periodicity of some properties of the f-element series[1, 10, 21, 23, 24, 27, 51, 61, 63, 65)].

The analogy schematically described in Fig. 8 is the basis for the construction of periodic tables of both series of f-electron elements. This is illustrated by the experimental findings that changes in some lanthanide and actinide properties recur regularly. Two types of such periodic tables can be compiled. The first one is as follows:

	0	I	II	III	IV	V	VI	VII
1. f^q	f^0	f^1	f^2	f^3	f^4	f^5	f^6	f^7
2. f^{7+q}	f^7	f^8	f^9	f^{10}	f^{11}	f^{12}	f^{13}	f^{14}

In the case of trivalent lanthanides we can establish the following order

La	Ce	Pr	Nd	Pm	Sm	Eu	Gd
Gd	Tb	Dy	Ho	Er	Tm	Yb	Lu

and in the case of trivalent actinides we have

Ac	Th	Pa	U	Np	Pu	Am	Cm
Cm	Bk	Cf	Es	Fm	Md	No	Lw

In this form, the periodic table of f-elements consists of two periods and eight groups. In the case of trivalent lanthanides this accounts for the fact that the properties studied change in an analogous way in the two periods: La–Gd and Gd–Lu. This table is also consistent with the tendency of some lanthanides to appear in the +2 or +4 oxidation states.

According to that what was shown earlier, each of the two subgroups, f^0–f^7 and f^7–f^{14}, can be further divided into two smaller units, known as segments. Within the segments, each consisting of four elements, the changes in the free energy of extraction or complex formation, lattice parameters, redox potentials, and other properties are similar. This leads to the following form of the periodic table of f-elements:

	0	I	II	III
1. f^q	f^0	f^1	f^2	f^3
2. f^{4+q}	f^4	f^5	f^6	f^7
3. f^{7+q}	f^7	f^8	f^9	f^{10}
4. f^{11+q}	f^{11}	f^{12}	f^{13}	f^{14}

This table consists of four periods and four groups. It is interesting to note that in the 4 f series the zero and third group contain lanthanides in the + 3 oxidation state (except perhaps for Nd), whereas the lanthanides, which can also appear in + 4 or + 2 oxidation states occupy the first and second group. The main feature, however, is the analogy with the decrease of the differences in many experimental values of trivalent lanthanides which is, generally observed within each of the four segments. Thus, for trivalent lanthanides, the following periodic table can be compiled:

	0	I	II	III
1. f^2	La	Ce	Pr	Nd
2. f^{4+q}	Pm	Sm	Eu	Gd
3. f^{7+q}	Gd	Tb	Dy	Ho
4. f^{11+q}	Er	Tm	Yb	Lu

The largest difference is between the elements forming the 0 and I groups, i.e. between La and Ce, Pm and Sm, Gd and Tb, Er and Tm, in the consecutive periods of the above table, respectively. The smallest difference is between the elements forming the II and III groups, i.e. between Pr and Nd, Eu and Gd, Dy and Ho, Yb and Lu, in the consecutive four periods, respectively. Therefore pairs of lanthanides such as Pr–Nd, Eu–Gd, Dy–Ho, or Yb–Lu are extremely difficult to separate by techniques conserving the trivalent character of both elements. This also explains the comparatively high separation factors known for such pairs of trivalent lanthanides as La–Ce, Gd–Tb, Pm–Sm and Er–Tm. Hence, one can see the practical significance of the regularities, i.e. periodical changes in some properties, for the separation of neighboring f-elements.

The situation is more complicated for actinides because of the differentiation of stable oxidation states. According to the results discussed in Sect. 4 (see Fig. 6), the analogy with the changes in unit cell volumes observed between the second and third segment of trivalent actinide compounds can be assumed as the basis for the compilation of the following two periods of homologous actinides existing in the + 3 oxidation state.

	0	I	II	III
2.	Np	Pu	Am	Cm
3.	Cm	Bk	Cf	Es

The results for tetravalent actinide compounds, which represent the analogy between the first and second segment can be treated similarly. In this case, the following two periods can be introduced:

	0	I	II	III
1.	Th	Pa	U	Np
2.	Pu	Am	Cm	Bk

There are very few hexavalent actinide. They contain only one segment, from U to Am. One can expect, however, that if the heavier hexavalent actinides were experimentally available they would show same homologous features within the following arrangement.

	0	I	II	III
1.	U	Np	Pu	Am
2.	Cm	Bk	Cf	Es
3.	Es	Fm	Md	No
4.	Lw			

Similarly, it can be expected that in the tetravalent state, Bk and Md can be homologs of Th and Pu, as well as Cf and No, those of Pa and Am, and so on. Also, in the case of the trivalent state, the elements of the fourth period Fm, Md, No, and Lw can exhibit some similarities to the 0, I, II and III group, respectively.

6 Application of the Double-Double Effect to the Solution of the Problem of the Inner Coordination Sphere of Aquoions

Due to the work performed both in our and other laboratories, the existence of the double-double effect in the properties mentioned above is firmly established. Because of that and because of the physical meaning of the double-double effect established by Jørgensen, it can be used for studying some problems of the coordination chemistry of f-electron elements[10, 22]. Provided other factors do not disturb, one may expect the double-double effect to appear in all properties which depend in a simple way on f-element radii and ground-term stabilization energies. Because there is a correlation with the nephelauxetic effect, the bond covalency in f-element compounds can be estimated on the basis of the double-double effect[10, 57]. This effect can be applied as a very sensitive indicator of changes in the coordination number of lanthanide and actinide ions in solutions. This has been utilized for solving the problem of the first coordination sphere of aquoions of trivalent lanthanides which, in turn, helps to solve the problem occurring with trivalent actinides[56–58].

The problem of the first hydration sphere of lanthanide(III) ions has been discussed for many years. Among the hypotheses which have been advanced, the most convincing was that of Spedding et al.[80–83]. According to their hypothesis, the coordination number in aquoions of lanthanides(III) decreases from 9 to 8, somewhere between Nd and Tb[81]. The reversal of the trend in the thermodynamic parameters of many reactions involving lanthanide aquoions, e.g. enthalpy of extraction and complexation [10, 65, 84] or enthalpy of crystallization of the ethyl sulfates[85], has been a good confirmation of the change in the inner sphere water coordination of lanthanide ions. On the other hand, Geier et al. have claimed that the coordination number of the aquoions of lanthanides remains constant along the whole series[86, 87]. This view has been supported by relatively few authors[88–92]. It was shown, however, by Staveley et al.[85] that in the case of the enthalpy of crystallization of the ethyl sulfates, a monotonic variation with Z can occur after elimination of the effect of variable hydration of the simple lanthanide ions, using a thermodynamic cycle. Such elimination can be achieved through complexation of the lanthanide ions with certain ligands such as dipicolonate or diglycolate.

We have attempted to use isomorphic ethyl sulfates of lanthanides to study the coordination number of lanthanide aquoions on the basis of the double-double effect. This idea arose immediately after we realized that the data of Staveley and coworkers show the double-double effect in the enthalpy of complexation after elimination of the enthalpies of hydration[84]. It was assumed that the coordination number of lanthanides aquoions can be determined from the thermodynamic quantities of the solubility of the isomorphic ethyl sulfates $[Ln(H_2O)_9](C_2H_5SO_4)_3$ because in these compounds the lanthanide ions are surrounded in the crystal lattice by 9 water molecules in the known arrangement: face-centered trigonal prism (Helmholz structure[93]), symmetry point group D_{3h}. Therefore, the change in the coordination number of lanthanides in their aquoions should be reflected by the variation of the free energy and enthalpy of solubility with the atomic number Z. Instead of the solubility, the cocrystallization of microamounts of lanthanides, and also some actinides, with the ethyl sulfate matrix $[Ln(H_2O)_9](C_2H_5SO_4)_3$ has been studied[56–58]. It was shown that the thermodynamic quantities of truly isomorphic homogeneous cocrystallization with the lanthanide ethyl sulfates are closely related to those of solubility[94]. The results which have been obtained are as follows:

1. The free energy of cocrystallization decreases from La to Pm, then increases slowly from Pm to Gd and rather sharply from Gd to Lu. Because of the constancy of the coordination number in the solid phase with respect to water molecules such a trend suggests the change in the hydration number of lanthanides, occurring between Pm and Gd.

2. In the Gd–Lu range half of the double-double effect is observed. It should be noted that in this case the curvatures of segments are directed upward, as is true for plots of stability constants or relative extraction coefficients, i.e. quantities increasing with Z[10, 65] (see Sect. 5).

3. There is no indication of the existence of the effect in the La–Pm range. The fact that the effect does not appear in this range together with that the effect does occur in the Gd–Lu range can, in our opinion, be regarded as evidence of the change in the coordination number of lanthanide aquoions from 9 to 8 between Pm and Gd. This conclusion results from the following reasoning.

As it was explained in Sect. 2 the magnitude of the double-double effect depends on the change of Racah's parameters under the nephelauxetic action of final ligands. The greater the distance between the initial and final ligands located in the nephelauxetic series, the stronger the double-double effect. The effect can therefore be observed if the ligands in the initial lanthanide compound differ from those in the final compound with respect to the position in the nephelauxetic series of ligands. The fact that no double-double effect is observed for light lanthanides (from La to Pm) shows that the electron-donor properties of water molecules relative to these lanthanides are the same both in the crystalline ethyl sulfates and respective aquoions. It also reveals that the number and arrangement of water molecules in these aquoions are the same as in the crystalline ethyl sulfates, i.e. the coordination number is equal to 9 and the Helmholz structure for the arrangement can be assumed. The appearance of the double-double effect in the Gd–Lu range means that the electron-donor ability of coordinating water molecules in aquoions of heavy lanthanides is different from that in crystalline ethyl sulfates. The increase in the electron-donor ability can be caused by the decrease in the distances between the central lanthanide ion and the coordinating water molecules. Such a decrease, in turn, can be due to the reduction in the coordination number of aquoions from 9 to, most probably, 8. The eight water molecules in the square antiprism arrangement (symmetry point group D_{4d} or D_{2d}) are, on an average, closer than the nine water molecules in the tricapped trigonal prism arrangement. In summary, these results show that the coordination number of lanthanide aquoions is equal to 9 in the La–Pm range; it is equal to 8 in the Gd–Lu range and intermediate (a mixture) between Pm and Gd. This example demonstrates 1. that the double-double effect can be applied to the solution of a complicated basic problem of the coordination chemistry of f-electron elements, 2. that, as established by Jørgensen, the physical meaning of the double-double effect has been confirmed experimentally, and 3. that the problem of the innersphere coordination of the lanthanide aquoions seems to be solved definitely, although indirectly.

Another possibility is to apply the double-double effect to the differentiation of outer- from inner-sphere complexes of lanthanides and actinides because both the magnitude and direction of the effect depend on the difference in the electron-donor ability between water in the aquoion and the ligand in the complex. Therefore, a substantial difference between outer- and inner-sphere complexes with respect to the double-double effect should exist. Research in this field, using lanthanide complexes with thiocyanates as an example, is being made in our laboratory.

It should be mentioned that although the problem of the change in the innersphere coordination of aquoions of the lanthanide series was solved by means of the double-double effect in 1973[55–57], the work in this field is still continued by our group[95, 96] as well as by others[92, 97–103].

The idea of one of the authors to employ the double-double effect to the determination of the coordination number of the inner transition elements in solutions[57] has been developped in his recent works[95, 96]. Thus, the double-double effect has proved to be a useful instrument to solve some difficult problems of the f-element coordination chemistry when other methods are often deceiving.

7 References

1. Fidelis, I., Siekierski, S.: J. Inorg. Nucl. Chem. *28*, 185 (1966)
2. Fidelis, I.: Bull. Acad. Polon. Sci. Ser. Sci. Chim. *18*, 681 (1970)
3. Fidelis, I.: Siekierski, S.: J. Inorg. Nucl. Chem. *33*, 3191 (1971)
4. Peppard, D. F., Mason, G. W., Lewey, S.: J. Inorg. Nucl. Chem. *31*, 2271 (1969)
5. Peppard, D. F. et al.: J. Inorg. Nucl. Chem. *32*, 339 (1970)
6. Fidelis, I.: J. Inorg. Nucl. Chem. *32*, 997 (1970)
7. Cunningham, B. B.: Pure Appl. Chem. *27*, 43 (1971)
8. Janowski, A., Lewandowski, W.: Proc. 15th Internat Conf. Coord. Chem. Moscow 1973
9. Hubert, S., Hussonnois, M., Guillaumont, R.: J. Inorg. Nucl. Chem. *35*, 2923 (1973)
10. Fidelis, I.: Double-Double Effect and its Contribution to the Lanthanide Chemistry, Report INR-1449/V/C/A, Warsaw 1973
11. Fidelis, I., Siekierski, S.: 10th Rare Earth Res. Conf., Carefree, Arizona, Proc. *2*, 919 (1973)
12. Alstad, J., Augustson, J. H., Farbu, L.: J. Inorg. Nucl. Chem. *36*, 899 (1974)
13. Guillaumont R., David, F.: Radiochem. Radianal. Lett. *17*, 25 (1974)
14. Fidelis, I., Krejzler, J.: J. Radioanal. Chem. *31*, 45 (1976)
15. Lewandowski, W.: Doctoral Dissertation, University of Warsaw, Warsaw 1978
16. Fidelis, I., Siekierski, S.: J. Inorg. Nucl. Chem. *29*, 2629 (1967)

17. Moeller, T.: Inorg. Chem. Series One (ed.) Bagnall, K. W., Vol. 7, University Park Press, Baltimore 1972
18. Cleverty, A. Mc.: Inorganic Chemistry of the Transition Elements (ed.), Chapt. 4, p. 376, 1972
19. Gmelins Handbuch der Anorganischen Chemie, Vol. *8*, Transurane, Part A, Die Elemente, 1973
20. Jørgensen, C. K.: Angew. Chem. *85*, 1 (1973)
21. Fidelis, I., Siekierski, S.: Periodicity of Some Properties of Lanthanides and Actinides, Section Lecture on 13th Internat. Conf. on Coord. Chem., Krakow-Zakopane, Sept. 1970
22. Fidelis, I., Siekierski, S.: The Double-Double Effect and its Significance for the Chemistry of Lanthanides and Actinides, Proc. 1st Scientific Conf. Iraq Atomic Energy Commission, Bagdad, April 1975
23. Siekierski, S.: The Double-Double Effect – Experimental Evidence and Significance, EUCHEM Conf. Chem. Rare Earths, Helsinki 1978
24. Fidelis, I.: 2nd Internat. Congr. Phosphorus Compounds, Boston, Mass. USA, April 1980
25. Keller, C.: The Chemistry of the Transuranium Elements, p. 80, Verlag Chemie 1971
26. Müller, W., Maas, K.: "Themen zur Chemie der Lanthanide und Actinide", Heidelberg 1974
27. Siekierski, S., Fidelis, I.: Extraction Chromatography (Eds.) T. Braun, G. Ghersini, Akademiai Kiadó, Budapest 1975
28. Keenan, T. K., Asprey, L. B.: Inorg. Chem. *8*, 235 (1969)
29. Zachariasen, W. H.: Acta Crystallogr. *2*, 388 (1949)
30. Asprey, L. B., Kruse, F. H., Penneman, R. A.: Inorg. Chem. *6*, 544 (1967)
31. Shankar, J., Khubchandani, P. G., Padmanabhan, V. M.: Anal. Chem. *29*, 1374 (1957)
32. Larson, A. C., Roof, R. B., Jr., Cromer, D. T.: Acta Crystallogr. *17*, 555 (1964)
33. Asprey, L. B.: J. Am. Chem. Soc. *76*, 2019 (1954)
34. Asprey, L. B. et al.: J. Am. Chem. Soc. *79*, 5825 (1957)
35. Asprey, L. B., Keenan, T. K.: Inorg. Nucl. Chem. Lett. *4*, 537 (1968)
36. Jørgensen, C. K.: J. Inorg. Nucl. Chem. *32*, 3127 (1970)
37. Jørgensen, C. K.: Orbitals in Atoms and Molecules, Academic Press 1962
38. Jørgensen, C. K.: Oxidation Numbers and Oxidation States, Springer Verlag, Berlin, Heidelberg, New York 1969
39. Jørgensen, C. K.: Modern Aspects of Ligand Field Theory, North-Holland Publ. Co., Amsterdam 1971
40. Jørgensen, C. K.: Struct. Bonding *13*, 199 (1973)
41. Jørgensen, C. K.: Progr. Inorg. Chem. *4*, 73 (1962)
42. Jørgensen, C. K.: Acta Chem. Scand. *17*, 251 (1963)
43. Jørgensen, C. K., Pappalardo, R., Schmidtke, H. H.: J. Chem. Phys. *39*, 1422 (1963)
44. Jørgensen, C. K., Pappalardo, R., Rittershaus, E.: Z. Naturf. *A 19*, 424 (1964)
45. Jørgensen, C. K., Pappalardo, R., Rittershaus, E.: Z. Naturf. *A 20*, 54 (1965)
46. Jørgensen, C. K.: Proc. 5th Rare Earth Res. Conf., Ames, Iowa 1965
47. Jørgensen, C. K.: Helv. Chim. Acta *50*, 131 (1967)
48. Jørgensen: C. K.: Chem. Phys. Lett. *2*, 549 (1968)
49. Jørgensen, C. K.: Progr. Inorg. chem. *12*, 101 (1970)
50. Nugent, C. J.: J. Inorg. Nucl. Chem. *32*, 3485 (1970)
51. Siekierski, S.: J. Inorg. Nucl. Chem. *32*, 519 (1970)
52. Harmon, H. D. et al.: J. Inorg. Nucl. Chem. *34*, 1381 (1972)
53. Hussonnois, M. et al.: Radiochem. Radioanal. Lett. *15*, 47 (1973)
54. Hubert, S. et al.: J. Inorg. Nucl. Chem. *36*, 1361 (1974)
55. Mioduski, T., Siekierski,S.: Proc. 15th Internat. Conf. Coord. Chem., Moscow, June 1973, *2*, 514 (1973)
56. Mioduski, T., Siekierski, S.: J. Inorg. Nucl. Chem. *37*, 1647 (1975)
57. Mioduski, T.: Doctoral Dissertation, Institute of Nuclear Research of Warsaw, Warsaw 1973
58. Mioduski, T., Siekierski,S.: J. Inorg. Nucl. Chem. *38*, 1989 (1976)
59. Guillaumont, R.: PTCHEM Conf. Warsaw 1975
60. Pyykkö, P.: EUCHEM Conf. Helsinki 1978
61. Siekierski, S., Fidelis, I.: J. Inorg. Nucl. Chem. *34* 2225 (1972)
62. Siekierski, Fidelis, I.: 24th IUPAC Congress, Hamburg 1973
63. Siekierski,S.: J. Inorg. Nucl. Chem. *33*, 377 (1971)

64. Zachariasen, W. H.: 4th Internat. Transplutonium Element Symp. Baden-Baden, Germany, Sept. 1975
65. Fidelis, I.: Doctoral Dissertation, Institute of Nuclear Research of Warsaw, Warsaw 1967
66. Sinha, S. P.: Helv. Chim. Acta *58*, 1978 (1975)
67. Sinha, S. P.: Structure and Bonding *30*, 1 (1976)
68. Fidelis, I.: Inorg. Nucl. Chem. Lett. *12*, 475 (1976)
69. Fidelis, I.: Nukleonika *12*, 477 (1967)
70. Fidelis, I.: Helv. Chim. Acta *62*, 2046 (1979)
71. Bommer, H.: Z. Anorg. Allg. Chem. *241*, 273 (1939)
72. Kettelle, B. H., Boyd, G. E.: J. Am. Chem. Soc. *69*, 2800 (1,947)
73. Thompson, S. G., Cunningham, B. B., Seaborg, G. T.: J. Am. Chem. Soc. *72*, 2798 (1950)
74. Higgins, G. H., Street, K., Jr.: J. Am. Chem. Soc. *72*, 5321 (1950)
75. Kettelle, B. H., Boyd, G. E.: J. Am. Chem. Soc. *73*, 1862 (1951)
76. Wheelwright, E. J., Spedding, F. H., Schwarzenbach, G.: J. Am. Chem. Soc. *75*, 4196 (1953)
77. Mayer, S. W., Freiling, E. C.: J. Am. Chem. Soc. *75*, 5647 (1953)
78. Peppard, D. F.: 16th Internat. Congr. Pure Appl. Chem., Paris 1957
79. Hesford, E., Jackson, E. E., McKay, H. A. C.: J. Inorg. Nucl. Chem. *9*, 279 (1959)
80. Spedding, F. H. Atkinson, G.: The Structure of Electrolytic Solutions (Ed.) W. J. Hamer, Chap. 22, John wiley, New York 1959
81. Spedding, F. H., Pikal, M. J., Ayers, B. O.: J. Phys. Chem. *70*, 2440 (1966)
82. Spedding, F. H., Jones, K. C.: J. Phys. Chem. *70*, 2450 (1966)
83. Hale, C. F., Spedding, F. H.: J. Phys. Chem. *76*, 2925 (1972)
84. Fidelis, I.: Bull. Acad. Polon. Sci., Ser. Sci. Chim. *20*, 605 (1972)
85. Staveley, L. A. K., Markham, D. R., Jones, M. R.: J. Inorg. Nucl. Chem. *30*, 231 (1968)
86. Geier, G., Karlen, U., Zelewsky, A. v.: Helv. chim. Acta *52*, 1967 (1969)
87. Geier, G., Karlen, U.: Helv. Chim. Acta *54*, 1351 (1971)
88. Reuben, J., Fiat, D.: J. Chem. Phys. *51*, 4909 (1969)
89. Albertsson, J.: Acta Chem. Scand. *26*, 1023 (1972)
90. Grenthe, I., Hessler, G., Ots, H.: Acta Chem. Scand. *27*, 2543 (1973)
91. Anderegg, G., private communication
92. Grenthe, I.: Kemia-Kemi *5*, 234 (1978)
93. Helmholz, L.: J. Am. Chem. Soc. *61*, 1544 (1939)
94. Mioduski, T.: J. Radioanal. Chem. *31*, 139 (1976)
95. Mioduski, T.: J. Radioanal. Chem. *53*, 25 (1979)
96. Mioduski, T.: J. Radioanal. Chem. *53*, 37 (1979)
97. Spedding, F. H., Cullen, P. F., Habenschuss, A.: J. Phys. Chem. *78*, 1106 (1974)
98. Spedding, F. H., Rard, J. A.: J. Phys. Chem. *78*, 1435 (1974)
99. Spedding, F. H. et al.: J. Phys. Chem. *79*, 1087 (1975)
100. Spedding, F. H., Rard, J. A., Habenschuss, A.: J. Phys. Chem. *81*, 1069 (1977)
101. Habenschuss, A., Spedding, F. H.: Cryst. Struct. Commun. *7*, 535 (1978)
102. Habenschuss, A., Spedding, F. H.: J. Chem. Phys. *70*, 2797 (1979)
103. Habenschuss, A., Spedding, F. H.: J. Chem. Phys. *70*, 3758 (1979)

Chemical Luminescence Analysis of Inorganic Substances

Alla P. Golovina, Valentin K. Runov, and Nikita B. Zorov

Department of Chemistry, Moscow State University, 117234 Moskow, USSR

Besides low detection limits and high sensitivity, chemical luminescence analysis is relatively simple and requires inexpensive equipment. It is widely used in detecting trace amounts of an especially pure component in a mixture, in quantitative analysis of pure metals and alloys, semiconductor materials, soil, air, biological, and other specimens.

I. Introduction

Chemical luminescence analysis of inorganic compounds consists of various methods of detecting elements with the help of the luminescence phenomenon. These methods are based on chemical reactions which generally proceed in solutions and are accompanied with emission of radiation (rarely with quenching of this emission) at room or lower temperatures (often, at liquid nitrogen temperature, i.e. 77 K)[1]. Ultraviolet or visible light is usually used as the excitation source.

Both organic and inorganic compounds can serve as luminescence reagents. The majority of the luminescence techniques have been designed for the detection of cations, and a few for the detection of anions. These methods are highly sensitive and have low detection limits (10^{-1} – 10^{-4} μg/ml). Certain methods, however, exhibit low selectivity, especially those based on the use of the fluorescence of complexes of several cations with organic reagents. This shortcoming can be overcome by using appropriate masking reagents or by preliminary separation. A combination of different methods like fluorescence and solvent extraction or fluorescence and chromatographic techniques seems a rather promising route to enhance the selectivity.

Besides low detection limits and high sensitivity, chemical luminescence analysis is relatively simple and requires inexpensive equipment. It is widely used in detecting trace amounts of an especially pure component in a mixture, in quantitative analysis of mineral samples or their technological products, in the analysis of pure metals and alloys, semiconductor materials, soil, air, biological, and other specimens[1–24].

Rapid advances in chemical luminescence analysis of inorganic compounds can be illustrated by the following data. In 1958 about seventy fluorescence reactions were known[25], while in the period from 1969 to 1979 their number multiplied almost by a factor of seven[15–23]. Several monographs on luminescence analysis have been published. From time to time reviews on the topic appear in Analytical Chemistry and in Zavodskaya Laboratoriya (Industrial Laboratory) (in Russian). In the last decade numerous books have been devoted to a study of the theoretical principles of photoluminescence[26–40]. Some of them[27–36] have been translated into Russian[41, 42]. The books[43–47] have been published in the USSR, and the eminent monograph "Fotonika molekul kracitelei i rodstvennykh organicheskikh soedinenii" by A. N. Terenin[48] (Photonics of molecules of dyes and related organic compounds) has not lost its significance even now. The books[10, 13] on the use photoluminescence in the analysis of inorganic substances have been translated into Russian[49, 50] by "Mir" publishers. Some sections in "Analiticheskaya Khimiya Elementov" (Analytical Chemistry of Elements) issued by the V. I. Vernadsky Institute of Geochemistry and Analytical Chemistry and in "Analiticheskie Reagenty" (Analytical Reagents) published in collaboration with the USSR Academy of Sciences (since 1973) deal with the problems in luminescence analysis of inorganic compounds.

II. Theoretical Principles of Chemical Luminescence Analysis

Chemical luminescence analysis is an important application of the luminescence phenomenon in practice. Naturally, the development of new luminescence techniques and the improvement of the existing methods are based on preliminary investigations of the luminescence phenomenon per se. Therefore, at present, great attention is paid to the study of the theoretical principles of luminescence analysis which we shall briefly outline in the following pages.

A. Classification of Luminescence[6)]

There is no generally accepted classification of luminescence analysis methods. As a rule, these techniques are classified according to the method or source of excitation, and by the mechanism or kinetics of the luminescence process.

Depending on the excitation method used, luminescence techniques are divided into photoluminescence excited by photons, cathodoluminescence generated under the action of cathode rays, X-ray luminescence excited by X-rays, candoluminescence generated under the action of heat, and sonoluminescence excited by ultrasound. Emission generated under the action of a stream of ions from alkali metals in vaccum is called ionoluminescence; radiation which atoms emit on optical excitation in plasma is known as atomic fluorescence; chemiluminescence is the emission of radiation generated by the energy of chemical reactions, it does not require an external excitation source. The excitation source needed in each particular case is chosen on the basis of this classification.

Thus, most of the analytical luminescence methods employing the radiation emission from metal ion complexes with organic reagents are carried out in solutions. Usually, UV or visible light is used to excite this type of luminescence, while UV, cathode or X-rays are applied to excite phosphors in solid crystals (crystallophosphors).

Photoluminescence is widely used in chemical luminescence analysis, on which greater attention is focussed in this review. Here, we shall not touch upon the works dealing with X-ray fluorescence, atomic fluorescence and other types of luminescence: a comprehensive list of literature on these topics is given in recent reviews[20, 51, 52)].

In the other classification based on the mechanism of the luminescence process, we distinguish resonance, spontaneous, stimulated, and recombination luminescence. Resonance luminescence is exhibited by atoms and certain simple molecules in the gaseous phase. R. Wood was the first to observe the emission of radiation from metallic sodium atoms. When excited by light of wavelength $\lambda = 588$ nm, sodium atoms are promoted to an excited state. When they return to their normal state, they emit luminescence quanta equal to the absorbed quanta. This emission of radiation, called resonance luminescence, is rarely observed. As a rule, excited atoms lose a certain amount of energy, as a result of which the irradiated quanta have less energy than the absorbed quanta, and the wavelength of fluorescenced radiation is longer than that of the absorbed radiation. At present, luminescence of gaseous atoms is known as atomic fluorescence which has become an independent area in luminescence analysis[52)]. Spontaneous luminescence is the short-time emission of radiation which occurs spontaneously at room temperature. Stimulated luminescence is the persistent fluorescence generally observed at low temeratures or in rigid media (polymer films, glassy media, solid adsorbents). Both these types of radiation are characteristic of molecular systems (complex organic molecules, their complexes with metal ions and certain inorganic compounds with a molecular crystalline lattice). Therefore, spontaneous and stimulated radiation are usually called molecular luminescence. Of course, recombination luminescence is a much more complicated phenomenon. This type of luminescence is observed when radicals or ions recombine to yield excited molecules (say, the emission of radiation from luminol). It may also occur in different gases and especially in phosphors in solid crystals (recombination luminophores)[53–56].

B. Molecular and Recombination Luminescence[6, 10, 36, 48]

Most of the luminescence methods are based on the formation of luminescent complexes of metal ions with organic reagents which irradiate molecular fluorescence. Let us consider a simple scheme for the molecular fluorescence of complicated organic molecules many of which are luminescence reagents. It should be noted that complicated organic molecules may exist either in the singlet (S) or triplet (T) state. In the singlet state the spins of two electrons in a molecule are antiparallel while in the triplet state they are parallel. The ground state of organic molecules with an even number of electrons (i.e. molecules but not radicals) is, according to Pauli's principle, singlet (S_0). If in the transition to the excited state, the spin of one of the electrons does not change (i.e. the multiplicity does not change), this state will also be singlet (S_1, S_2, . . . , etc). In the triplet state, molecules possess two electrons with parallel spins, hence, on being excited, one of the electrons changes the direction of its spin, and the multiplicity changes as well. Owing to electrostatic repulsion of two electrons, the energy of a triplet state is lower than that of a singlet state[27]. Absorption of radiation usually involves raising of a molecule from one of the several vibrational levels in the ground electronic state to different vibrational levels of excited states (for simplicity, only one of the transitions is shown in Fig. 1). This process occurs in about 10^{-15} s. An excited molecule may lose the absorbed energy in the form of luminiscence (radiative loss) or lose it due to vibrational relaxation or radiationless transition.

Vibrational relaxation (vr) or vibrational deactivation of singlet and triplet levels is determined by collisions during which excess vibrational and rotatinal energies gained by a molecule in the cours of excitation is converted into kinetic energy of molecules. In solution, this conversion takes place rapidly with a rate constant $k_{vr} \sim 10^{13}s^{-1}$. Consequently, molecules relax to the lowest vibrational level of the corresponding electronically excited state (i.e. without any change in the electronic energy). In Fig. 1 vibrational relaxation is shown by vertical wavy lines.

Raditionless transitions occur between isoenergy (degenerate) vibration-rotational levels of different electronic states. Since the total energy of a system does not change here, no radiation is emitted. These processes are shown in Fig. 1 by horizontal wavy lines. Here, the most rapid process is the internal conversion (ic) – radiationless transition between states of identical multiplicity (for example, $S_1 \rightsquigarrow S_0$, $T_2 \rightsquigarrow T_1$, Fig. 1). Its rate constant is quite high: $k_{ic} \sim 10^{12}s^{-1}$, which is the reason for the extremely weak emission from higher electronically excited states. On the other hand, internal conversion from S_1 to S_0 occurs so slowly that fluorescence can successfully compete with this conversion. Intersystem crossing (isc) is the radiationless transition between states of different multiplicities (for instance, $S_1 \rightsquigarrow T_2$, $S_1 \rightsquigarrow T_1$, $T_1 \rightsquigarrow S_0$, $T_1 \rightsquigarrow S_1$, Fig. 1). Intersystem crossing from S_1 is a rather rapid process ($k_{isc} \sim 10^9s^{-1}$), so that it may compete with fluorescence and reduce its quantum yield. Radiationless deactivation of

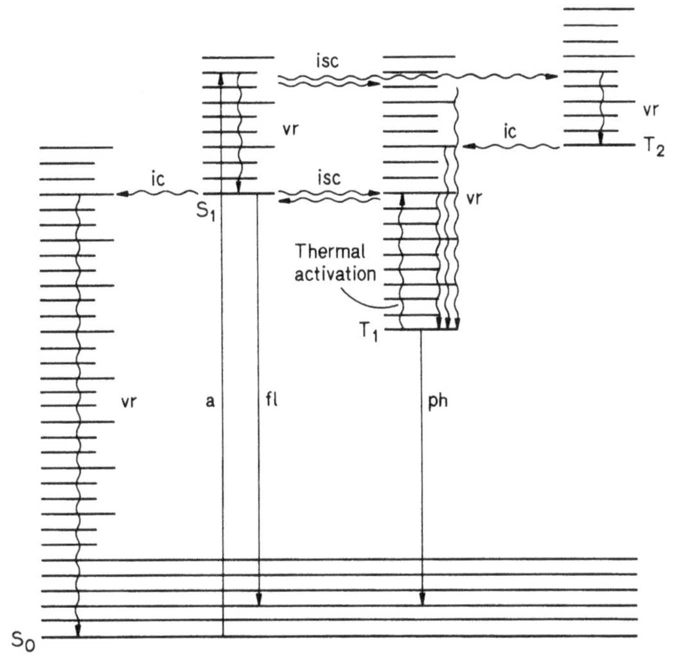

Fig. 1. Diagramme of radiative (a – absorption, fl – fluorescence, ph – phosphorescence), radiationless (ic – internal conversion, isc – intersystem crossing) transitions and processes of vibrational relaxation (vr) in organic molecules

the lowest triplet state ($T_1 \rightsquigarrow S_0$) takes place at a slower rate ($k_{isc} \sim 10^{-2} - 10^4 s^{-1}$) and can compete with phosphorescence. In order that the transition $T_1 \rightsquigarrow S_1$ may occur, it is necessary that the triplet molecules have to be thermally activated to a vibrational level equal in energy to the lowest vibrational level of S_1. This process is responsible for one of the mechanisms of delayed fluorescence. Since delayed fluorescence has not been applied so far in the analysis of inorganic substances, we shall not consider it in discussing radiative transitions (vertical solid lines in Fig.1).

Fluorescence (fl) takes place when transitions occur between states of the same multiplicity. It is a rapid process ($k_{fl} \sim 10^6 - 10^9 s^{-1}$). For the luminescent reagents used in analytical chemistry, fluorescence corresponds to the transition $S_1 \rightarrow S_0$. As a result of internal conversions and radiationless transitions, fluorescence occurs from the lowest vibrational level of S_1 to the vibrational levels of S_0.

Phosphorescence (ph) occurs when transitions take place between states of different multiplicities. In analytical chemistry, only the transition $T_1 \rightarrow S_0$ (phosphorescence) is considered. The spectrum of phosphorescence is shifted towards longer wavelengths than the spectrum of fluorescence. Since the transition $T_1 \rightarrow S_0$ is forbidden with regard to spin, the rate constant of phosphorescence ($k_{ph} \sim 10^{-2} - 10^4 s^{-1}$) is considerably smaller than that of fluorescence. Having a long radiative lifetime, triplet molecules can readily lose energy in various radiationless processes. Therefore, phosphorescence is usually observed at low temperatures or in rigid media.

From the foregoing, it is evident that the two luminiscence phenomena – fluorescence and phosphorescence – are competetive processes. Besides, they also compete with radiationless loss of energy of an excited molecule. Which of these processes will dominate depends on the type of a molecule, its environment and experimental conditions.

Recombination luminiscence is used, of course, to a limited extent, in luminiscence analysis of inorganic substances to detect trace amounts of a large number of elements. Recombination crystallophosphors, for example sulfides, are ionic crystals consisting of matrix (CaS, SrS, ZnS, CdS and others) whose crystalline lattice may easily suffer distortion on the introduction of minute amounts of metal ions (like Ag^{+}, Cu^{2+}, Mn^{2+} etc.) called activators. The mechanism of this luminiscence is explained in terms of the band theory of solids[54–56].

As a result of interactions with other atoms, the electronic levels of certain atoms in the crystalline lattice of solids are split into several sublevels, depending on the interacting atoms. The set of such sublevels forms energy bands the most important of which determining the optical and electrical properties of a crystal, are valence and conduction bands. The valence band formed by the energy levels of anions of a crystal is populated with electrons. The conduction band consisting of cation levels is not filled with electrons. These two bands are partitioned by an energy gap called the forbidden band (Fig. 2). When excited, electrons of a crystallophosphor may pass from the valence band into the conduction band. Since this band is the generalized level of a crystal as a whole, electrons can move freely over this band. The exciting energy should be greater than the energy corresponding to the forbidden band width (> ΔE, Fig. 2). For several crystallophosphors, the forbidden band width is about 2–10 eV. Therefore, UV light can easily be used to excite these crystals. Activators play an essential role in luminescence because they constitute a part of the luminescence centre of the crystallophosphor. Like ions of the matrix, also the activators have their energy levels in the forbidden band. Their normal level is, as a rule, slightly above the valence band (a) whereas the excited level (a') slightly below the bottom of the conduction band (Fig. 2). In the presence of foreign ions or other defects of the crystalline lattice of matrix, favourable conditions are created for the formation of energy levels capable of trapping and retaining for some time the electrons which fall into the conduction band. These levels are therefore called

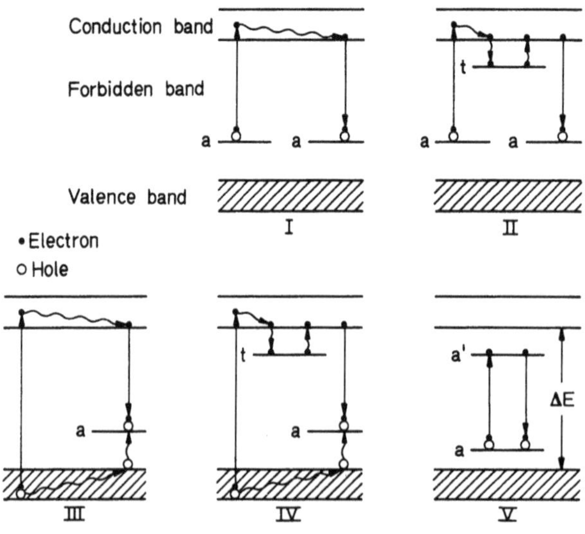

Fig. 2. Luminescence of crystallophosphors excited through the luminescence centre (I, II), through the basis of a crystal (III, IV) or when excitation and emission of radiation are associated with the activator (V)

capture levels or electron traps (t). They are located in the forbidden band below the bottom of the conduction band (Fig. 2). When excited, the electron is raised from the activator level through the luminescence centre to the conduction band, moves freely over it, and passes into the level of some other excited luminescence centre, recombines with it and emits light(I). This process is rather rapid ($\tau \sim 10^{-8} - 10^{-9}$s). If an electron is captured by traps before it recombines, then the process (II) may be persistent ($\tau \sim 10^{-4}$ s or even longer). When the basis of a luminescent crystal is excited, the electron passes into the conduction band. Thus, in place of this electron, a "hole" is formed which exhibits the properties of a positive charge. Radiation takes place as a result of recombination of an electron (coming from the conduction band) with the hole at the activtor level (III). This luminescence is momentary. As in case II, if an electron is captured at traps, luminescence may be persistent (IV).

Extensive use is made of characteristic luminophores in analytical chemistry. Examples of such luminophores are crystallophosphors, activated mercury-like ions, uranyl ions, and lanthanides. They exhibit a glow similar to molecular luminescence. Their absorption and emission of radiation are associated with the electronic transitions occurring within a luminescence centre – the activator (Fig. 2 V).

C. Basic Characteristics of Luminescent Substances

The important characteristics of luminescent substances that determine their use in luminescence analysis are the eletronic spectra of absorption, excitation, luminescence, luminescence yield, and luminescence persistence.

1. Electronic Spectra of Absorption, Excitation and Luminescence

The spectra of luminescent substances reflect the energy transitions between the ground and excited states of molecules, atoms or ions. Transitions from the ground to excited states are characterised by the absorption spectrum, while the reverse transitions by the spectra of luminescence fluorescence and phosphorescence (Fig. 1). Excitation spectra, unlike the absorption spectra, determine only the region of active absorption of energy and show the dependence of the luminescence intensity on the wavelength of the exciting light. As a rule, in luminescence analysis such a wavelength of exciting light is chosen for which maximum luminescence intensity is observed. Excitation spectra are especially needed in studying frozen solid solutions of organic compounds, their complexes with metal ions or halide complexes of mercury-like ions and powdery crystallophosphors. For these substances, it is extremely difficult to measure the absorption spectra due to strong scattering and reflection of the exciting light.

2. Energy and Quantum Yield of Luminescence

An important characteristic of luminescent substances is the luminescence yield which shows how effectively is the exciting light transformed into luminescence. Generally, a distinction is made between energy and quantum yields. The energy yield (φ_E) is determined as the ratio of the energy (E_e) emitted by a substance to the absorbed excitation energy (E_a):

$$\varphi_E = E_e/E_a \tag{1}$$

The quantum yield (φ) is defined as the ratio between the rates of emission of radiation and absorption of ligth. It is easy to derive a relationship between these two quantities:

$$\varphi_E = E_e/E_a = h\bar{\nu}_e N_e/(h\bar{\nu}_a N_a) = \varphi\bar{\nu}_e/\bar{\nu}_a \tag{2}$$

where $\bar{\nu}_a$ and $\bar{\nu}_e$ are the mean effective frequencies of absorbed and emitted ligth, respectively. Usually, $\bar{\nu}_a > \bar{\nu}_e$ and $\varphi_E < \varphi$.

S. I. Vailov was the first to suggest a method[57, 58] for determining the energy and quantum yields of fluorescence (φ_{fl}) for several substances like fluorescein, rhodamine, quinine sulfate. They have eventually become the standards with respect to which the yields of other substances are measured. Relative methods of determining φ_{fl} for various substances, including complexes of metal ions with organic and inorganic reagents, are described in[33, 59–61]. The quantity φ_{fl} is one of the parameters characterizing the sensitivity of the fluorometric method. The higher the value of φ_{fl}, other conditions being the same, the more sensitive is the method.

A decrease in φ_{fl} is called quenching of fluorescence. Quenching may result both from internal factors (radiationless processes which transfer a molecule into the state S_0) as well as from external factors (interactions of excited molecules with other molecules). In the latter case, we distinguish photophysical and photochemical mechanisms of quenching. In analytical chemistry we have, first of all, take the concentration of a fluorescent substance into account, because the so-called concentration quenching may set in at higher concentrations. This phenomenon commences at a certain "threshold" concentration (C_0) and the yield is an exponential function of concentration:

$$\varphi_{fl} = \varphi_0 e^{-K(C-C_0)} \tag{3}$$

where φ_0 is the yield of fluorescence at infinite dilution and φ_{fl} = const, if $C \leqslant C_0$.

The magnitude of "threshold" concentration C_0 and the constant K are specific to each fluorescent substance. The "threshold" concentration is determined from the concentration dependence of the luminescence intensity. In dilute solutions, the fluorescence intensity (I) is a linear function of concentration:

$$I = 2{,}3\, I_0 \varepsilon C l \varphi_{fl} \tag{4}$$

where I_0 is the intensity of exciting light, ε the molar absorptivity, l the length of the sample cell. Temperature affects the magnitude of φ_{fl}: frequently, the fluorescence intensity decreases with increasing temperature. In[61–64] this phenomenon is attributed to the increasing rate of intersystem crossing from the S_1-state with rising temperature. Fluorescence intensity increases perceptibly at lower temperatures. A similar picture is observed if the substance is enclosed in a rigid medium.

3. *Lifetime of Excited States*

The lifetime (τ) of excited states is also an important characteristic of luminescent substances. It is much shorter for molecular fluorescence than for recombination luminescence. Quenching of these two types of luminescence also takes place via different laws. Thus, quenching of molecular luminescence proceeds according to the exponential law:

$$I = I_0 e^{-t/\tau} \tag{5}$$

where I and I_0 are the luminescence intensities at instant t and at the moment when excitation ceases (t = 0), respectively; τ is the lifetime of excited states of molecules. Quenching of recombination luminescence takes place according to the hyperbolic law:

$$I = A\,(b+t)^{-\beta} \tag{6}$$

where A, b, β are constants typical of a substance; generally, $\beta < 2$. Methods of determining the lifetime of excited states of molecules are described in detail in[10, 60, 65].

D. *Kinetics of Photoluminescence* [10, 36, 66]

A knowledge of the basic characteristics of luminscent substances permits the determination of the rate constants of photophysical processes discussed in Sect. B.

If photochemical reactions and intersystem crossings do not take place in the upper singlet states, then the general kinetic pattern of the photoprocesses occurring during absorption of light quanta can be expressed as follows:

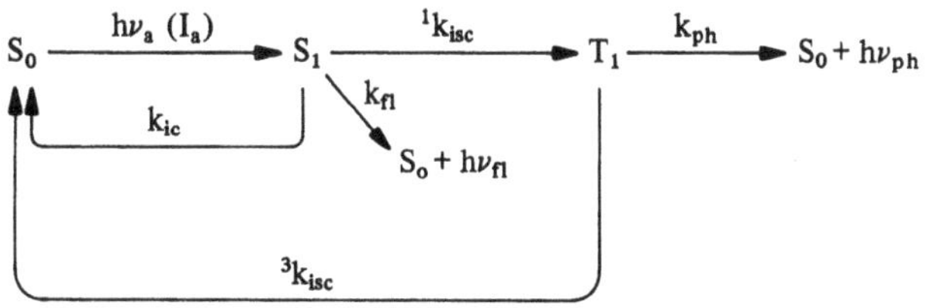

(7)

where I_a is the rate of absorption of light by the state S_0, $^1k_{isc}$ and $^3k_{isc}$ are the rate constants of intersystem crossings $S_1 \rightsquigarrow T_1$ and $T_1 \rightsquigarrow S_0$, respectively, k_{ic} is the rate constant of internal conversion $S_1 \rightsquigarrow S_0$, k_{fl} and k_{ph} are the rate constants of radiative transitions of fluorescence and phosphorescence, respectively. Assuming for the excited state the following steady-state scheme, the rates of formation and deexcitation of the state S_1 are equal:

$$I_a = \Sigma^1 k[S_1] \tag{8}$$

where

$$\Sigma^1 k = k_{fl} + k_{ic} + {}^1k_{isc}$$

and

$$\varphi_{fl} = \frac{k_{fl}[S_1]}{I_a} \tag{9}$$

Using Eq. (8), we can write

$$\varphi_{fl} = \frac{k_{fl}}{\Sigma^1 k} \tag{10}$$

Hence, the true (experimental) lifetime (τ_{fl}) of the state S_1 is

$$\tau_{fl} = 1/\Sigma^1 k \tag{11}$$

Therefore

$$k_{fl} = \varphi_{fl}/\tau_{fl} \tag{12}$$

The rate constant of fluorescence k_{fl} can likewise be calculated from the absorption spectra of fluorescent compounds[67–69]. The radiative lifetime of fluorescence, τ_0, i.e. the liefetime of the state S_1, when the latter vanishes due solely to emission of fluorescence, is equal to

$$\tau_0 = 1/k_{fl} \tag{13}$$

Then

$$\varphi_{fl} = \tau_{fl}/\tau_0 \tag{14}$$

For triplet molecules, the steady-state approximation gives

$${}^1k_{isc}[S_1] = \Sigma^3 k[T_1] \tag{15}$$

where

$$\Sigma^3 k = k_{ph} + {}^3k_{isc}$$

The quantum yield of phosphorescence (φ_{ph}) is

$$\varphi_{ph} = \frac{k_{ph}[T_1]}{I_a} \tag{16}$$

and the radiative lifetime of phosphorescence (τ_{ph})

$$\tau_{ph} = 1/\Sigma^3 k \tag{17}$$

The expressions for the quantum yield of triplet formation (φ_T)

$$\varphi_T = \frac{\text{Rate of formation of } T_1}{\text{Rate of absorption of light by } S_0} \tag{18}$$

and the quantum efficiency of phosphorescence (Θ_{ph})

$$\Theta_{ph} = \frac{\text{Rate of emission of light from } T_1}{\text{Rate of formation of } T_1} \tag{19}$$

can, by virtue of Eq. (15), be written as

$$\varphi_T = {}^1k_{isc}[S_1]/I_a = \Sigma^3k[T_1]/I_a \tag{20}$$

$$\Theta_{ph} = k_{ph}[T_1]/\Sigma^3k[T_1] \tag{21}$$

Then, using Eqs. (20, 21), the expression (16) can be rewritten as

$$\varphi_{ph} = k_{ph}[T_1]/I_a = (k_{ph}/\Sigma^3k)({}^1k_{isc}[S_1]I_a) = \Theta_{ph}\varphi_T \tag{22}$$

The value of ${}^1k_{isc}$ is calculated from experimental data:

$$^1k_{isc} = \varphi_T/\tau_{fl} \tag{23}$$

The value of φ_T is determined using the method of calculating triplets[70], by means of flash photolysis[71–73], or quenching of fluorescence by heavy atoms[74], or by triplet transfer of energy[75, 76] techniques.

For many aromatic molecules it has been found[74, 77] that

$$\varphi_{fl} + \varphi_T \sim 1 \tag{24}$$

Hence

$$\varphi_T \sim 1 - \varphi_{fl} \tag{25}$$

Knowing Θ_{ph} and τ_{ph}, we can, in principle, calculate the value of k_{ph} with the help of the expression:

$$k_{ph} = \Theta_{ph}/\tau_{ph} \tag{26}$$

Since Θ_{ph} cannot be measured directly, usually the experimentally determinable quantity φ_{ph} is used instead:

$$\varphi_{ph}/k_{ph} = [T_1]/I_a = \Sigma^3k[T_1]/\Sigma^3kI_a = \tau_{ph}\varphi_T \tag{27}$$

Hence

$$k_{ph} = \varphi_{ph}/\varphi_T\tau_{ph} \tag{28}$$

Table 1 lists the experimental values of quantities and expressions needed in calculating the rate constants of certain photoprocesses.

Table 1. Experimental values of quantities and expressions needed in calculating the rate constants of certain photoprocesses

Constant	Quantity	Expression[a]
k_{fl}	$\varepsilon = f(\tilde{\nu})$	$k_{fl} = 2900\, n^2 \tilde{\nu}_a^2 \int \varepsilon d\tilde{\nu}$ (from Bowen E. J., and Wokes, F.[67])
		$k_{fl} = 2900\, n^2 \int \frac{(2\tilde{\nu}_0 - \tilde{\nu})^3}{\tilde{\nu}} \varepsilon d\tilde{\nu}$ (from Förster, T.[68])
		$k_{fl} = 2880\, \frac{n_{fl}^3}{n_a} \cdot \frac{\int I d\tilde{\nu}}{\int I \tilde{\nu}^{-3} d\tilde{\nu}} \cdot \int \frac{\varepsilon d\tilde{\nu}}{\tilde{\nu}}$ (from Birks, J. B., and Dyson, D. J.[69])
	φ_{fl}, τ_{fl}	$k_{fl} = \varphi_{fl}/\tau_{fl}$
$^1k_{isc}$	φ_T, τ_{fl}	$^1k_{isc} = \varphi_T/\tau_{fl}$
k_{ph}	$\varphi_{ph}, \varphi_T, \tau_{ph}$	$k_{ph} = \varphi_{ph}/\varphi_T \tau_{ph}$

[a] ε is the molar absorptivity, $\tilde{\nu}$ the wave number: $\tilde{\nu}_a$ – for the maximum of absorption band, $\tilde{\nu}_0$ – for the mirror image line; n is the refractive index of the solvent, n_a and n_{fl} are the average refractive indexes of the solvent in the absorption and fluorescence regions, respectively; I is the fluorescence intensity

E. Main Laws of Molecular Fluorescence[6, 48, 78, 79]

An important feature of molecular fluorescence lies in the mutual positions of the absorption and the fluorescence spectra.

1. Stokes-Lommel Law

D. Stokes[80] formulated a rule according to which fluorescence light has always a wavelength longer than that of the absorbed exciting ligth. For several complicated organic molecules and their complexes with metal ions, however, the spectra are partially overlapped and the fluorescence quanta in this range are greater than the absorbed quanta. This spectral range is called anti-Stokes region. It arises due to the specific amount of vibrational energy in absorbing molecules. The distance between the maxima in absorption and fluorescence spectra is called Stokes shift. E. Lommel[81] corrected the Stokes rule as follows: the fluorescence spectrum as a whole and its maximum are always shifted relative to the absorption spectrum and its maximum towards longer wavelengths. The Stokes-Lommel law is strictly satisfied for a wide range of fluorescent substances.

Anti-Stokes fluorescence depends on the concentration of thermally activated molecules in the ground state. Therefore, its intensity can be reduced by lowering the temperature. This fact is virtually not used in inorganic luminescence analysis. But in several cases of organic analysis, e.g. in the determination of anthracene in the presence of phenanthrene[10], it has been possible to suppress 90% of the luminescence intensity of the blank[1] due to anti-Stokes fluorescence of phenanthrene by decreasing the temperature.

1 The blank is a solution which contains everything in the sample solution except the substance to be determined. The blank signal is simply subtracted from the total analytical signal[13]

2. V. L. Levshin's Mirror Image Relationship[78)]

According to this rule the normalized spectra of absorption and fluorescence expressed as a function of the frequency are specularly symmetric with respect to a line normal to the frequency axis at the intersection of these two spectra. The mathematical expression for the mirror image relationship is

$$\nu_a + \nu_{fl} = 2\,\nu_0 \tag{29}$$

or

$$\Delta\nu = \nu_a - \nu_{fl} = 2(\nu_a - \nu_0) \tag{30}$$

where ν_0 is the frequency of the symmetry line corresponding to the frequency of pure electronic transition (the so-called 0–0 transition) ν_a and ν_{fl} are the symmetric frequencies of absorption and fluorescence, respectively. From Eq. (30) it is seen that if the mirror image relationship is strictly satisfied, the ν_a-dependence of $\Delta\nu$ is a straight line with a slope equal to 2.

3. B. I. Stepanov's Universal Relationship Between Absorption and Fluorescence Spectra[79)]

On the basis of general thermodynamical considerations, without any regard for the specific feature of a particular molecule, the universal relationship between the spectra of absorption and fluorescence can be expressed as

$$I_\nu/\varepsilon_\nu = D(T)\nu^3 e^{-h\nu/kT} \tag{31}$$

where I_ν is the rate of emission of fluorescence at the frequency ν, ε_ν the molar absorptivity at the frequency ν, D(T) a constant which depends on the excitation conditions and temperature T, and k the Boltzmann constant.

These laws are extremely helpful in luminescence analysis, especially in interpreting the spectra and in establishing the nature of energy levels of molecules. If either the mirror image or the universal relationship is strictly satisfied, then the shape of one of the spectra (absorption or fluorescene) can be determined from the shape of the other. The mirror image shows that the vibrational systems of the electron levels of S_0- and S_1-states have identical structures and can be used to evaluate their relative populations and relative probabilities of absorptive and radiative transitions as well as to determine the frequency of pure electron transitions.

S. I. Vavilov has established the relationships which express the fluorescence spectrum and yield as a function of the wavelength of the exciting light[57, 58)].

4. Excitation Wavelength Invariance of Fluorescence Spectra

Fluorescence spectra of several organic molecules and their complexes with metal ions in solutions do not depend on the wavelength of the excitation light, if the wavelength lies within the range of their electronic absorption spectral bands. This is attributed to the fact mentioned in Sect. B that radiation takes place from the lowest vibrational level of the state S_1.

5. S. I. Vavilov's Law

The energy yield of fluorescence increases proportional to the wavelength of the excitation ligth and then remains (in a certain range) constant, and thereafter decreases steeply in the region of overlap of absorption and fluorescence spectra (i.e. in the anti-Stokes region). The magnitude of φ_{fl} does not vary right up to the anti-Stokes region and then sharply decreases. S. I. Vavilov formulated his law as follows: fluorescence yield may remain constant if the wavelength of the excitation is converted on the average into a longer wavelength, and the yield decreases sharply if the excitation wavelength is converted into a shorter wavelength (anti-Stokes region).

These laws can be used in choosing a proper wavelength of the excitation light. Since the fluorescence yield does not depend on the excitation wavelength, and since the fluorescence intensity is determined by the molar absorptivity (see Eq. (4)) of the substance, sensitivity of determination can be enhanced by changing the excitation wavelength. For instance, the absorption spectra of dyes (rhodamine, acridine) and their compounds are characterized by low-intensity bands in the UV region and more intensive bands in the visible range. On excitation with visible ligth, the fluorescence intensity of these compounds increases considerably. Moreover, measuring the luminescence of rhodamines and other organic dyes becomes less expensive when using instruments with conventional ligth sources like tungsten filament lamps in place of expensive mercury or xenon arc lamps.

F. Metròlogical Characteristics in Luminescence Analysis[6, 13, 82–85)]

In presenting the results of luminescence analysis or in choosing or comparing different methods of determination of elements, the following metrological characteristics (figures of merit) are used: sensitivity, precision, accuracy, and detection limit. In the following pages we shall use the terminology approved by IUPAC and the Department of Analytical Chemistry, The USSR Academy of Sciences.

From the calibration curve – the plot of analytical signal (y) luminescence intensity vs. analyte concentration C in a series of standards having known analyte concentration – the sensitivity (H) is expressed as the first derivative of the calibration function at a given concentration C_i:

$$H = \left(\frac{\delta y}{\delta C}\right)_{C_i} \qquad (32)$$

If the calibration curve is linear over the whole range of analyte concentration, the sensitivity is equal to $\Delta y/\Delta C$ and it remains constant in this concentration range.

Precision is the degree to which the measured values of luminescence intensity or analyte concentration are close to each other. Precision is characterized by relative standard deviation S_r:

$$S_r = \sigma/\bar{y} \tag{33}$$

where σ is the standard deviation:

$$\sigma = \sqrt{\sum_{i=1}^{n} \frac{(y_i - \bar{y})^2}{n - 1}} \tag{34}$$

y_i are the measured values of luminescence, $\bar{y}$ is the mean of a series of measurements. If $n < 20$, the symbol σ is used for S. The relative standard deviation S_r in this case can be written as

$$S_r = S/\bar{y} \tag{35}$$

Accuracy is a measure to judge how close the difference between the average and actual analyte concentration is to zero.

The detection limit ($C_{min,\,p}$) is the smallest analyte concentration which can be detected by a given method with a reasonable certainity level P. This concentration is given by

$$C_{min,\,p} = \frac{y_a - \bar{y}_0}{H} \tag{36}$$

where y_a is the limiting detectable luminescence intensity which can still be recorded, $\bar{y}_0$ the average luminescence intensity of the blank:

$$y_a = \bar{y}_0 + KS_0 \tag{37}$$

where K is the protection factor, S_0 the standard deviation of measurements. Usually, K is taken to be 3. Then we have

$$C_{min,\,p} = 3\,S_0/H \tag{38}$$

In other words, the detection limit is that analyte concentration leading to a luminescence intensity which is three times the background standard deviation. To evaluate $C_{min,\,p}$ at least twenty y_0 measurements should be used. If 3 S is chosen, then for a Gaussian distribution the confidence level is $P = 0.997$. At low analyte concentrations, however, more likely are broader and asymmetrical distributions. In general, 3 S corresponds to a confidence level P of about 0.9. Thus, the detection limit is not directly measured but determined by extrapolating the measurement made at higher analyte concentrations. Therefore, its numerical magnitude is only a rough estimate ($S_r \sim 0.5$). It is believed[13] that if the ratio of the detection limits $\leqslant 2$ then these limits almost do not differ from each other.

In presenting the analytical results, the confidence interval should necessarily be mentioned, i.e. the interval in which the average concentration C exists at a given confidence level. For a Gaussian distribution of n parallel measurements ($n > 20$) and probability $P = 0.95$, the confidence interval is given by the expression

$$\bar{C} \pm 2\sigma/\sqrt{n} \qquad (39)$$

In practice, the number of measurements n is usually less than 20. In such cases, the confidence interval is taken to be

$$\bar{C} \pm \frac{t_{p,n}S}{\sqrt{n}} \qquad (40)$$

where $t_{p,n}$ is the Student t-value which depends on P and n.

III. Luminescence Determination of Elements

At present, luminescence determination methods are available for almost all elements of the periodic table (Fig. 3). Most extensively used are methods whereby fluorescent, more rarely phosphorescent, complexes of elements with organic ligands are obtained. Many methods utilize the native luminescence of lanthanoides (III), uranyl, mercury-like and other ions in crystallophosphors and complexes with inorganic and organic ligands and also chemiluminescence.

A. Determination of Elements by Luminescence of Their Complexes with Organic Ligands

Most luminescence methods use organic reagents. This section briefly deals with the luminescence and electronic structure of both the organic reagents and their complexes with the metal ions to be determined.

1. Luminescence and Electronic Structure of Organic Molecules

Spectral and luminescence properties of organic molecules have been found to depend on the electronic structure, or the relative position of lower electronically excited states, whatever their orbital nature and multiplicity that change under the influence of various structural factors and intermolecular interactions[86–89]. In terms of the characteristics of electronic orbitals of organic molecules which are used in analytical chemistry the four most important types can be indicated (in the S_0-state):

I. $\sigma\pi$ (aromatic hydrocarbons);
II. $\sigma\pi l$ (compounds with –N<, –O–, –S– etc. groups);

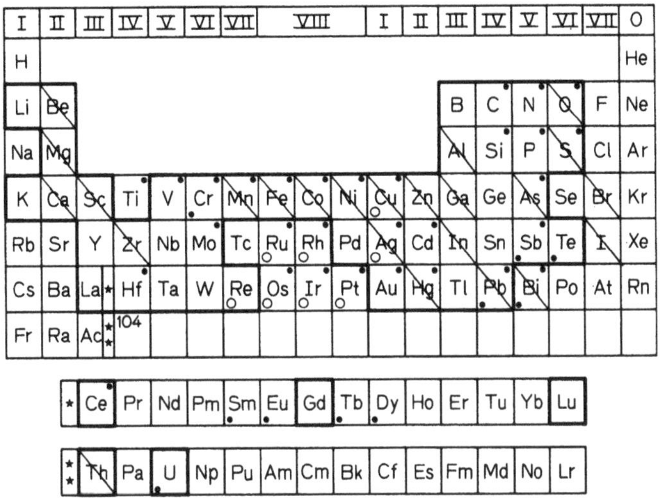

Fig. 3. Luminescence methods of determining elements developed in the recent decade. Fluorescence of complexes with organic ligands – □; native luminescence of complexes with inorganic and organic ligands (proceeding of determination are developed – □, possibility of determination are mentioned – □); chemiluminescence – □; titration with luminescence indicators – ⧅

III. $\sigma\pi n$ (compounds with >C=O, $-NO_2$, –N=O, >C=N–, >C=S etc. groups);

IV. $\sigma\pi ln$ (compounds with –N<, –O–, –S– etc. groups and >C=O, $-NO_2$, –N=O, >C=N–, >C=S etc. groups).

π-Orbitals are significantly delocalized in comparison with σ-orbitals. n-Orbitals localized on the heteroatom are in the molecular plane and orthogonal to π-orbitals. The contribution of the l-orbital to π-delocalization depends on the turning angle of the $2p_z$-orbital axis with respect to the neighboring π-orbital axis of the aromatic ring. Since the l-orbital is occupied by a pair of electrons, its contribution is chiefly information of the S_0-state[87)]. In a later paper it has been shown that l-orbitals can also significantly contribute to the excited states of organic molecules[90)]. Since the electronic state is dictated by multiplicity and the nature of the electronic transition which is responsible for it, the following electronic transitions for the four types of molecules are possible: $\pi \rightarrow \pi^*$, $\pi l \rightarrow \pi^*$, and $n \rightarrow \pi^*$ and the associated $S_{\pi\pi^*}$, $T_{\pi\pi^*}$, $S_{\pi l\pi^*}$, $T_{\pi l\pi^*}$, $S_{n\pi^*}$, and $T_{n\pi^*}$ lower excited states.

An important step in the systematization of the spectral-luminescence properties of organic molecules was the determination of the dependence of the S_1-T_1-splitting on the orbital nature of the states[27, 44, 86–89, 91–95)]. Because its value depends on the overlap of orbitals, S_1-T_1-splitting is small for $n\pi^*$-states (2000–4000 cm^{-1})and for $\pi\pi^*$-states it is equal to 6000–12 000 cm^{-1}[27, 44, 89)]. The intermediate orbital properties of $\pi l\pi^*$-states in comparison with $\pi\pi^*$- and $n\pi^*$-states result in S_1-T_1-splitting remaining within the range of 4000–8000 cm^{-1}[89)]. Consequently, the value of this splitting is an important characteristic of the orbital nature of the states and an indication of the electronic structure of the molecules.

Another important stage in systematizing the organic molecules were investigations into the fine structure of their absorption spectra, fluorescence and phosphorescence at 77 and 4.2 K[44, 89, 92, 93, 96–107].

Orbital classification of molecules helped D. N. Shigorin to systematize molecules in terms of their spectral-luminescence properties[86–89]. Now spectral luminescence groups

Table 2. Spectral-Luminescence groups and types of organic molecules and some of their characteristics (Molar absorptivity of the longest wavelength band – ε_{max}; Rate constants of fluorescence – k_{fl}, phosphorescence – k_{ph}, intersystem crossing – k_{isc}; Lifetimes of fluorescence – τ_{fl}, phosphorescence – τ_{ph})

Molecules		Spectral-luminescence groups of molecules				
Types	Orbitals	I	II	III	IV	V

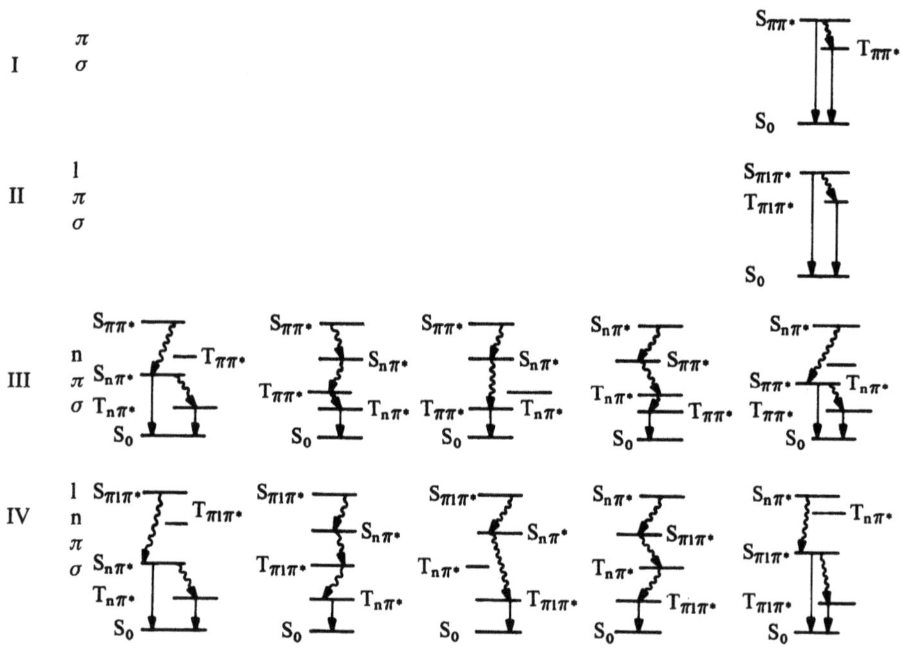

Spectral-luminescence characteristics	I	II	III	IV	V
ε_{max}	1–10^3	1–10^3	1–10^3	10^3–10^5	10^3–10^5
	(if transition is not symmetrically forbidden)				
k_{fl}, s^{-1}	$S_{n\pi^*} \rightarrow S_0$ 10^6				$S_{\pi\pi^*} \rightarrow S_0$ $(\pi l\pi^*)$ 10^7–10^9
k_{ph}, s^{-1}	$T_{n\pi^*} \rightarrow S_0$ 10^2–10^4	$T_{n\pi^*} \rightarrow S_0$ 10^2–10^4	$T_{\pi\pi^*} \rightarrow S_0$ $(\pi l\pi^*)$ 10^{-2}–10	$T_{\pi\pi^*} \rightarrow S_0$ $(\pi l\pi^*)$ 10^{-2}–10	$T_{\pi\pi^*} \rightarrow S_0$ $(\pi l\pi^*)$ 10^{-2}–10
k_{isc}, s^{-1}	$S_{n\pi^*} \rightsquigarrow T_{n\pi^*}$ 10^6–10^8	$S_{n\pi^*} \rightsquigarrow T_{\pi\pi^*}$ $(\pi l\pi^*)$ 10^9–10^{10}	$S_{n\pi^*} \rightsquigarrow T_{\pi\pi^*}$ $(\pi l\pi^*)$ 10^9–10^{10}	$S_{\pi\pi^*} \rightsquigarrow T_{n\pi^*}$ $(\pi l\pi^*)$ 10^9–10^{10}	$S_{\pi\pi^*} \rightsquigarrow T_{\pi\pi^*}$ $(\pi l\pi^*)$ $(\pi l\pi^*)$ 10^6–10^8
τ_{fl}, s	10^{-6}–10^{-7}				10^{-7}–10^{-10}
τ_{ph}, s	10^{-4}–10^{-2}	10^{-4}–10^{-2}	10^{-1}–10^2	10^{-1}–10^2	10^{-1}–10^2

(vertical columns, Table 2) are added to the types mentioned above (horizontal rows, Table 2). Molecules of the same group display qualitatively identical spectral-luminescence properties. Molecules of group I characterize low-intensity $n\pi^*$-fluorescence and short-lived $n\pi^*$-phosphorescence. For molecules of group II, only short-lived phosphorescence from the $n\pi^*$-state and for molecules of groups III and IV, long-lived phosphorescence from $\pi\pi^*$- or $\pi l\pi^*$-states are observed. Molecules of group V exhibit both fluorescence and long-lived phosphorescence from the $\pi\pi^*$- or $\pi l\pi^*$-states. If in frozen solutions the radiation occurs from $\pi\pi^*$-states, then progression of carbon skeleton frequencies is characteristic of the vibrational structure of fluorescence and phosphorescene spectra and for $n\pi^*$-phosphorescence it is dominated by a progression of vibration frequencies of the atomic group which is responsible for the $n \rightarrow \pi^*$-transition[108–114]. The fluorescence and phosphorescence spectra from $\pi l\pi^*$-states have, as a rule, a low resolution vibrational structure[115–117].

Electronic absorption spectra of various spectral-luminescence groups of organic molecules depend on the solvent and on the acidity of the medium. Shifts of absorption spectra are attributed to changes in the donor – acceptor interaction in the S_0- and S_1-states of the molcule[10, 36, 48]. In the replacement of a nonpolar solvent by a polar one, the long wavelength absorption bands are responsible for $\pi \rightarrow \pi^*$-transitions and $\pi l \rightarrow \pi$-transitions are usually shifted to longer wavelengths whereas the absorption bands of $n \rightarrow \pi^*$-transitions undergo hypsochromic shifts. The shift of $n\pi^*$-absorption bands to the region of short wavelengths in polar solutions is explained by the formation of an intermolecular hydrogen bond between molecules of the solvent and the dissolved species, the bond in the S_0-state being usually stronger than in the S_1-state. This results in increased energy of the $n \rightarrow \pi^*$-electronic transition. These effects are not, however, unambiguous indications of the precise nature of electronic transitions[118, 119]. Molecule protonization has a strong impact on $n \rightarrow \pi^*$- and $\pi l \rightarrow \pi^*$-transitions[10, 36, 48]. Furthermore, if the S_1-state is a weaker proton acceptor than the S_0-state, then protonization results in a hypsochromic shift of the longer wavelength absorption band. This is observed, for instance, for aromatic amines ($\pi l \rightarrow \pi^*$-transitions). On the other hand, protonization of N-heterocyclic bases leads to a bathochromic shift of lower wavelength absorption bands because molecules of these compounds bind the proton in the S_1-state stronger than in the S_0-state.

Consequently, to classify an organic molecule with a spectral-luminescence group and to select the best luminescence reagent it is necessary to (1) determine from absorption and luminescence spectra the energies of lower electronic excited states; (2) measure the lifetimes and quantum yields of fluorescence and phosphorescence; (3) investigate the vibrational structure of absorption and emission bands; and (4) study the effect of solvent and acidity of the medium on the absorption spectra.

It should be noted that there are no pure $n\pi^*$-, $\pi\pi^*$- or $\pi l\pi^*$-states in organic molecules with n-, π- and l-electrons. What actually occurs is mixing of $n\pi^*$-, $\pi\pi^*$-, and $\pi l\pi^*$-states of the same multiplicity through electron vibrational interaction and in the case of different multiplicities, through spin-orbital intraction. This mixing is strongly dependent on the energy difference between these states.

2. *Luminescence and Electronic Structure of Metal Ion Complexes with Organic Ligands*

The relation of spectral-luminescence properties of metal ion complexes with organic ligands and their electronic structure has been studied much less extensively. Various views on the causes of the appearance, decay or change of the luminescence color in complexation have been reported[6].

Today, it is generally believed that spectral-luminescence properties of metal ion complexes with organic ligands are a function of the nature and relative position of the lower electronically excited states of the organic ligand of the complex and the metal ion that form from associated molecular and atomic orbitals[6]. Fig. 4 outlines the relative position of lower electronically excited levels and the most probable transitions in such complexes. Thus, if the electronic level of a metal ion (M_1) is above the S_1-level of the complex (Fig. 4A), then molecular luminescence (fluorescence or, more rarely, phosphorescence) is possible. In this case, the metal ion behaves as an inert atom which retains different parts of the organic molecule and is responsible for the formation of an additional ring. This is characteristic of complexes of nontransitional metal ions. Fluorescence spectra of such compounds are broad bands, sometimes with an obscured vibrational fine structure. Usually, they are shifted to the long wavelength region in comparison with the fluorescence spectra of the reagents. In some complexes of Be(II), Sc(III), Gd(III), La(III) etc. a decrease of temperature causes phosphorescence. The resultant spectra of the complexes undergo a bathochromic shift, in contrast to their fluorescence spectra. In this case, the metal ion acts as a disturbance factor which increases the probability of $S_1 \rightsquigarrow T_1$-intersystem crossing in the organic part of the complex. Complexes in which the electronic level of the metal ion is between the S_1- and T_1-levels (Fig. 4B) are non-fluorescent because of radiationless $S_1 \rightsquigarrow T_1$ and $S_1 \rightsquigarrow M_1$-transi-

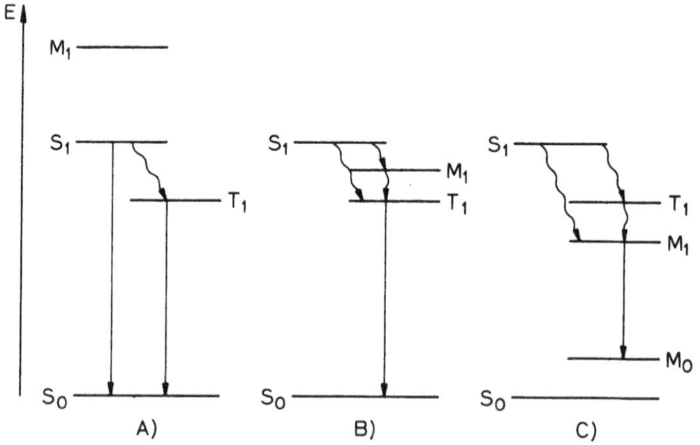

Fig. 4. Relative position of lower excited singlet state (S_1), triplet state (T_1) and metal ions (M_1), electronic levels, and transitions between them in metal ion complexes with organic ligands, (A) luminescence (fluorescence and phosphorescence) with levels of metal ion located above S_1-levels, (B) phosphorescence with metal ion levels located between S_1- and T_1-levels, (C) sensitized (native) luminescence with metal ion levels located below T_1-levels

tions. However, with decreasing temperature these complexes can display phosphorescence caused by the $T_1 \rightarrow S_0$-transition. This is characteristic of transitional metal complexes. If, however, the electronic level of the metal ion in the complexes is above the T_1-level (Fig. 4C), then radiationless energy transfer of electronic excitation from the S_1- and T_1-levels can occur: $S_1 \rightsquigarrow M_1$ and $T_1 \rightsquigarrow M_1$, and also ionic luminescence ($M_1 \rightarrow M_0$), usually referred to as sensitized luminescence, is observed. Such a luminescence is exhibited by complexes of Sm(III), Eu(III), Tb(III), Dy(III), Cr(III), and some platinum elements with organic ligands.

The classification of molecules according to their spectral-luminescence properties and the consideration of general laws of metal ion complex luminescence suggest ways for the purposeful search for luminescence reagents in the determination of inorganic ions. The applicability of organic luminescence reagents depends on the difference in the quantum yield values and positions of the complex and reagent luminescence spectra.

3. *Fluorimetric Determination of Elements*

The luminescence of chelate complexes with such widely known organic reagents as 8-hydroxyquinoline, hydroxyflavones (flavonols), hydroxyanthraquinones, hydroxyazo compounds, etc. and also ionic associates of aminoxanthenes, acrydines, and hydroxyxanthenes is more commonly used for the fluorimetric analysis of elements. The chelate-forming fluorimetric reagents can be organic reagents of all the five spectral luminescence groups. The most promising reagents are, however, nonfluorescent molecules of groups I–IV. The lack of fluorescence is attributable, as noted above, to the involvement of n-electrons in the formation fo the S_1-state. Formation of complexes is believed[10] to result in increased energy of the $n\pi^*$-state in comparison with the $\pi\pi^*$- or $\pi l\pi^*$-states and also in increased rigidity of the molecules, which leads to fluorescence. The value φ_{fl} of the resultant compounds is influenced by the metal ion. With 8-hydroxyquinolinates of nontransitional metals, as an example, their φ_{fl} at 77 K has been shown to be approximately equal to that of the organic reagent anion[120]. It is only for elements with large atomic numbers that the heavy atom effect can be observed which results in quenching of fluorescence. An analogous phenomenon is observed for complexes of Al(III), Ga(III), In(III), and Tl(III) with salicylidene-o-aminophenole[121]. The temperature decrease results in many cases in an increase of φ_{fl} and a decrease of the detection limit, for example, for Ga(III) and Nb(V) with lumogallione[122], Mg(II) with magnesone[122] and Be(II) with morin[123]. The research into the interdependence of the electronic structure and luminescence of the reagents and their metal ion complexes is being continued. The relations between the fluorescence characteristics of metal ion complexes and the structure of hydrazones[124–127], 8-hydroxyquinoline derivatives[128], azo-[129, 130] and hydroxyazomethine compounds[22] have been studied. Studies of 8-hydroxyquinolinates of nontransitional metals have led to the proposal that their fluorescence is caused by the increased conjugation degree of π-electrons as a result of disruption in the interaction of n-electrons of the nitrogen atom with the π-electronic system of the organic reagent molecule rather than by the formation of a new ring within the molecule of 8-hydroxyquinolinate[131].

It should be noted, however, that the spectral-luminescence properties of organic reagents of groups I–IV have not been studied in sufficient detail. Exceptions are only hydroxyanthraquinones[105, 132–134)], hydroxyazo-[135, 136)] and hydroxyazomethine compounds[137–139)].

For fluorimetric determinations of inorganic ions with molecules of spectral-luminescence group V, an additional condition is the separation of the reagent from the complex (in analytical chemistry solvent extraction is most often used for separation). The most widely used molecules of that group are basic dyes, in particular aminoxanthenes (rhodamines) by means of which over 50 elements of the periodic system are determined. These methods utilize differences in the extraction ability of simple salts and ionic associates of rhodamines with metal halide acidocomplexes or with anionic complexes containing organic ligands, then, fluorescence measurements of the extracts are made. In the latter case, some ionic associates are non-fluorescent. Fluorescence quenching depends on the nature of the organic ligand, particularly on the positions of lower electronically excited states of the dye molecules and the ligand[140, 141)]. Flash photolysis and pulsed excitation of delayed fluorescence and phosphorescence have revealed that for the extraction-fluorimetric determination of elements as ionic associates of this type, organic ligands whose energies of T_1-levels are ~ 30000 and 17000 cm^{-1} should be chosen[142)].

In Table 3 are compiled data on recent fluorimetric determinations of elements[124, 126, 128, 140, 143–289)] and in Table 4 the fluorimetric reagents most widely used in inorganic analysis.

Table 3[a]. Fluorimetric determination of elements developed in the recent decade

Element	Reagent	Method of estimation[b]	Detection limit (μg/ml)	Reference
1	2	3	4	5
Ag	Eosine and 1,10-phenantroline	Extr	3×10^{-2}	143, 144
	Eosine and pyridine	Extr	8×10^{-2}	145
	3,4,5,6-Tetrachlorofluorescein and 1,10-phenantroline	Sol	5×10^{-2}	146
Al	3-Amino-5-sulfosalicylic acid	Sol	5×10^{-3}	147
	Campherol	Sol	1×10^{-3}	148
	2,4-Dihydroxybenzaldehyde semicarbazone	Sol	2×10^{-4}	149, 150
	2,4-Dioxo-4-(4-hydroxy-6-methyl-2-pyron-3-yl)butyric acid ethyl ester	Extr	5×10^{-1}	151
	p-Hydroxybenzoic acid, 2-hydroxy-1-naphthylhydrazide	Sol	2×10^{-4}	152, 153
	2-Hydroxy-1-naphthalenedehyde benzoylhydrazone	Sol	1×10^{-1}	154
	8-Hydroxyquinoline-5-sulfonic acid	Sol	4×10^{-3}	128
	Lumogallion	Sol	5×10^{-5}	155
	Mordant blue 31	Sol	4×10^{-3}	156
	Resorcylidene-o-aminophenol	Sol	1×10^{-3}	157
	2-(Salicylidene-amino)-phenylarsonic acid	Sol	1×10^{-1}	158
	Sulfonaphtholazoresorcin	Sol	5×10^{-4}	159
	Sulfophenylazochromotropic acid	Sol	2×10^{-2}	160
	Superchrome Garnet Y	Sol	8×10^{-4}	161
Au	Crystal violet, buthylrhodamine B	Extr	2×10^{-3}	162
	p-Dimethylaminobenzylidenerhodanine	Extr	2×10^{-2}	163

Table 3[a]. (continued)

Element	Reagent	Method of estimation[b]	Detection limit (μg/ml)	Reference
1	2	3	4	5
B	Chromotropic acid	Sol	1×10^{-2}	164
	Morin and oxalic acid	Sol	1×10^{-4}	165–168
	Quercetin and oxalic acid	Sol	3×10^{-4}	165
Be	3-Amino-5-sulfosalicylic acid	Sol	5×10^{-3}	169
	Arsenazo III	Sol	4×10^{-2}	170
	Chlorophosphonazo III	Sol	4×10^{-2}	170
	Dinaphthoylmethane	Precipitate	1×10^{-2}	171
	2,4-Dioxo-4-(4-hydroxy-6-methyl-2-pyrone-3-yl)butyric acid ethyl ester	Extr	5×10^{-1}	151
	2-Ethyl-5-hydroxy-7-methoxyisoflavone	Extr	1×10^{-3}	172
	2-Ethyl-5-hydroxy-3-methylchromone	Extr	1×10^{-2}	173
	1-Hydroxy-2-carboxyanthraquinone	Sol	3×10^{-2}	174
	2-Hydroxy-1-naphthyl-o-amino-phenylarsonic acid	Sol	2×10^{-4}	175
	3-Hydroxy-2-naphthoic acid	Extr	2×10^{-3}	176
	8-Hydroxyquinoline-5-sulfonic acid	Sol		177
	2-Hydroxy-N-salicylidene-5-sulfoaniline	Sol	4×10^{-4}	178
	1-Hydroxy-6-methylxanthrone	Extr	1×10^{-3}	179
	2-(2-hydroxyphenyl)pyridine	Sol	2×10^{-3}	180
	2-Quinizarinesulfonic acid	Sol	1×10^{-3}	181
	Resorcylidene-o-aminophenylarsonic acid	Sol	4×10^{-4}	182
	Resorcylidene-cysteine	Sol	8×10^{-3}	183
	2-(Salicylidene-amino)-phenylarsonic acid	Sol	9×10^{-4}	184, 185
		Extr		186
	2-(Salicylidene-amino)-3,5-dimethylphenylarsonic acid	Extr	1×10^{-3}	186
	Tetracycline	Sol	5×10^{-3}	187
Ca	5,7-Dinitro-8-hydroxyquinoline and rhodamine B	Extr	5×10^{-2}	188
	Glyoxal, bis-(4-hydroxybenzoylhydrazone)	Sol		126
Cd	Calceine	Sol	2×10^{-3}	189
	Dibromofluorescein and 1,10-phenanthroline	Extr	2×10^{-2}	190
	Eosine and 1,10-phenanthroline	Extr	5×10^{-3}	144, 191
	Eritrosine and 1,10-phenanthroline	Extr	2×10^{-2}	191
	Glyoxal, bis-(4-hydroxybenzoylhydrazone)	Sol		126
	2-(2-Hydroxy-1-naphthyl)-dithiocarbazic acid, 4-chlorbenzyl ester	Sol	1×10^{-1}	192
	8-Hydroxyquinoline-5-sulfonic acid	Sol	2×10^{-2}	128
	1-(8-Hydroxy-2-quinolyl)-3,5-	Sol	6×10^{-2}	193
	dimethylpyrazole	Extr	9×10^{-3}	193
	8-Quinolinethyole	Extr	5×10^{-2}	194
	8-Quinolyl dihydrophosphate	Sol	2	195
Ce	8-Hydroxyquinoline-5-sulfonic acid	Sol	1×10^{-2}	196
	Morin	Sol	1×10^{-1}	197
Co	Eosine and 1,10-phenanthroline	Extr	1×10^{-2}	144
	Eosine and 2-pyridinecarbaldehyde hydrazone	Extr	8×10^{-3}	198
Cr	Safranine T	Extr	1×10^{-2}	199

Table 3[a]. (continued)

Element	Reagent	Method of estimation[b]	Detection limit (μg/ml)	Reference
1	2	3	4	5
Cu	1,3-Benzenedicarboxylic acid	Sol	2×10^{-2}	200
	Eosine and 1,10-phenanthroline	Extr	1×10^{-2}	144
Fe	Eosine and 1,10-phenanthroline	Extr	1×10^{-1}	144
	Phthalic acid and amalgamated zinc	Sol	1×10^{-3}	201
Ga	Acridine orange	Extr	5×10^{-3}	202
	Alizarin red C	Sol	2×10^{-1}	203
	Campherol	Sol	2×10^{-3}	204
	4-Amino-2,4-dihydroxybenzyl-2,3-dimethyl-1-phenyl-5-pyrazolone	Sol	3×10^{-3}	205
	Dodecyllumogallion (in the surfactant micellar system polyethylene glycol monolauryl ether)	Sol	1×10^{-3}	206
		Extr		206
	Hexyllumogallion	Extr	1×10^{-3}	207
	2-Hydroxy-5-methylbenzaldehyde-4-aminoantipyrine	Sol	2×10^{-4}	208
	2-Hydroxy-1-naphthalenecarbaldehyde thiosemicarbazone	Sol		209
	8-Hydroxyquinoline-5-sulfonic acid	Sol	2×10^{-2}	128
	Mordant blue 31	Sol	4×10^{-2}	156
	Quercetin and antipyrine	Extr	1×10^{-2}	210
	Quercetin, 3′-glucozide	Sol	6×10^{-3}	211
	8-Quinolinethyol	Extr	1×10^{-1}	212
	Resorcylidene-4-aminoantipyrine	Sol	1×10^{-3}	213
	Resorcylidene-cysteine	Sol	7×10^{-3}	182
	Rhodamine 3 GO	Extr	2×10^{-2}	214
	Rhodamine 4 G	Extr	2×10^{-3}	214
	Salicylidene-4-aminoantipyrine	Sol	1×10^{-3}	213, 215
	Superchrome Garnet Y	Sol	2×10^{-2}	161
Gd	Salicylic acid and rhodamine B	Extr	3×10^{-1}	216
Ge	Alizarin complexon and rhodamine 6 G	Flotation	2×10^{-3}	217
	Quercetin	Sol	1×10^{-2}	218
Hf	Miricetin	Sol	9×10^{-3}	219
	Morin	Sol	8×10^{-3}	220
	Quercetin	Sol	1×10^{-2}	221, 222
	Quercetinsulfonic acid	Sol	2×10^{-2}	223
Hg	Crystal violet, butylrhodamine B	Extr	5×10^{-3}	224
	Eosine and 1,10-phenanthroline	Extr	5×10^{-2}	144
	Rhodamine B	Extr		225
	Thyamine	Sol	1×10^{-2}	226
		Extr	4×10^{-2}	227
In	8-Hydroxyquinoline and tetraphenyl borate	Extr	2×10^{-2}	228
	8-Hydroxyquinoline-5-sulfonic acid	Sol	2×10^{-2}	128
	Lumogallion	Extr	2×10^{-2}	229
	8-Quinolinethyol and CH_3COO^-	Extr	5×10^{-2}	230
	4-Sulfonaphthol-(1-azo-1′)-2′,4′-dihydroxybenzene	Sol	2×10^{-2}	231

Table 3[a]. (continued)

Element	Reagent	Method of estimation[b]	Detection limit (μg/ml)	Reference
1	2	3	4	5
K	Dibenzo-18-crown-6 and anilinonaphthalenesulfonate	Extr	1	232
La	Glyoxal, bis-(4-hydroxybenzoylhydrazone)	Sol		126
Li	5,7-Dibromo-8-hydroxyquinoline	Sol	1×10^{-1}	233
Lu	Morin and diantipyrylmethane	Extr	2×10^{-2}	234
Mg	3′,4′-Dimethoxy-3-hydroxyflavone	Sol	1×10^{-1}	235
	8-Hydroxyquinoline-5-sulfonic acid	Sol	4×10^{-3}	128
	Morin	Sol	1×10^{-1}	236
	5-Pyrazolone-(4-azo-2)-1-naphthol-4-sulfonic acid	Sol	5×10^{-3}	237
Mn	Eosine and 1,10-phenanthroline	Extr	1×10^{-1}	144
Mo	8-Hydroxyquinoline and tetraphenyl borate	Extr	2×10^{-2}	238
	Morin	FSol	1×10^{-1}	239
	Rhodamine B and SCN^-	Sol	2×10^{-3}	240
Nb	Lumogallion and H_2O_2	Sol		241
	Lumogallion and F^-	Sol		241
	Morin and H_2O_2	Sol	4×10^{-3}	242
	Quercetin and H_2O_2	Sol	9×10^{-2}	243
	Sulfonaphtholazoresorcin and H_2O_2	Sol	4×10^{-2}	244
Ni	Eosine and 1,10-phenanthroline	Extr	1×10^{-1}	144
Pb	Eosine and 1,10-phenanthroline	Extr	1×10^{-1}	144
	Lumogen aqueous-blue	Sol	5	245
Pd	Eosine and 2,2′-dipyridyl or 1,10-phenanthroline or pyridine	Extr	2×10^{-2}	246
	Erythrosine and 1,10-phenanthroline or pyridine	Extr	2×10^{-2}	246
Re	Acridine orange	Extr	5×10^{-3}	247
	Acriflavin	Extr	8×10^{-3}	247
	Phenosafranine	Extr	8×10^{-2}	248
	Rhodamine B	Extr	2×10^{-3}	249
	Safranine T	Extr	4×10^{-3}	250
Sb	Crystal violet, rhodamine 3B	Extr	3×10^{-3}	251
	3,4′,7-trihydroxyflavone	Sol	2×10^{-3}	252
Sc	Anisic acid, 2-hydroxy-1-(1-naphthyl)hydrazide	Sol	2×10^{-3}	253, 254
	(2,4-Dihydroxybenzaldehyde)semicarbazone	Sol	2×10^{-3}	178, 255
	Eosine and 1,10-phenanthroline	Extr	3×10^{-2}	256
	2-Ethyl-5-hydroxy-7-methoxyisoflavone	Extr		257
	2-Ethyl-5-hydroxy-3-methylchromine	Extr	1×10^{-2}	258
	Hydroxystilbenyl complexon	Sol	1×10^{-2}	259
	Mordant blue 31	Sol	4×10^{-2}	156
	Myricitin	Sol	7×10^{-3}	260
	Myricitin, 3′-glucoside	Sol	7×10^{-4}	260
	2-Phenylquinoline-4-carboxylic acid	Sol	2×10^{-3}	261
	2-Phenylquinoline-4-carboxylic acid and rhodamine B	Extr	7×10^{-3}	262

Table 3[a]. (continued)

Element	Reagent	Method of estimation[b]	Detection limit (μg/ml)	Reference
1	2	3	4	5
	Quercetin, 7-glucozide	Sol	6×10^{-3}	260
	Salicylic acid, phenyl ester	Sol	1×10^{-3}	263
	Salicylic acid and rhodamine B	Extr	7×10^{-3}	262
Sn	Crystal violet, rhodamine 3 B and cupferron	Extr	4×10^{-2}	264
	o-Hydroxyhydroquinonephthalein	Sol		265
	Morin and antipyrine	Extr	5×10^{-3}	266
	Rhodamine B	Extr	1×10^{-1}	267
	3,4′,7-Trihydroxyflavone	Sol	3×10^{-4}	268
Ta	Brilliant green, butylrhodamine B	Extr	2×10^{-2}	269
	Morin and H_2O_2	Extr	6×10^{-3}	270
Te	2,4,4-Trihydroxybenzophenone	Sol	1×10^{-3}	271
Th	5,7-Dinitro-8-hydroxyquinoline and rhodamine B	Extr	7×10^{-3}	140, 272
	Myricetin	Sol	4×10^{-2}	273
	3,4′,7-Trihydroxyflavone	Sol	1×10^{-3}	274
Tl	Acridine orange	Extr	3×10^{-3}	275
	Acridine yellow	Extr	3×10^{-3}	276
	Acriflavin	Extr	2×10^{-3}	277
	Crystal violet, butylrhodamine B	Extr	3×10^{-3}	278
U	5,7-Dinitro-8-hydroxyquinoline and rhodamine B	Extr	7×10^{-3}	140
V	Benzoic acid and amalgamated zinc	Sol	5×10^{-4}	279
	1,4-Diamino-5-nitroanthraquinone	Sol	1×10^{-1}	280
	1,3-Dicarboxybenzoic acid	Sol	2×10^{-1}	199
W	Morin	Sol	3×10^{-3}	281
		FSol	7×10^{-2}	282
Y	5,7-Dinitro-8-hydroxyquinoline and rhodamine B	Extr	2×10^{-2}	140
	Methyl-bis(8-hydroxy-2-quinolyl)amine	Sol	2×10^{-2}	283
		Extr	1×10^{-2}	284
Zn	Benzimidazole-2-carbaldehyde-2-quinolylhydrazone	Sol	1×10^{-1}	124
	Dibromofluoresceine and 1,10-phenanthroline	Extr	1×10^{-2}	189
	Eosine and 1,10-phenanthroline	Extr	1×10^{-2}	144
	8-Hydroxyquinoline-5-sulfonic acid	Sol	2×10^{-2}	128
	8-Quinolyl, dihydrogen phosphate	Sol	1	194
	β-(Salicylidene-amino)ethanol	Sol	3×10^{-1}	285
Zr	3-Hydroxychromone	Sol		286
	β-Hydroxy-1-naphthyl-4-aminoantipyrine	Sol	1×10^{-1}	287
	8-Hydroxyquinoline	Extr	4×10^{-3}	288
	Myricetin	Sol	2×10^{-2}	218
	Quercetinsulfonic acid	Sol	4×10^{-2}	222
	3,4′,7-Trihydroxyflavone	Sol	2×10^{-3}	289

[a] Some data used from Refs. 1, 15

[b] Fluorescence of solutions (Sol), frozen solutions (FSol), extracts (Extr) of metal ion complexes

Table 4. Most widely used fluorimetric reagents in inorganic analysis

Element	Reagent
1	2
Ag	Eosine and 1,10-phenanthroline
	Butylrhodamine B
	Pyronine G
Al	8-Hydroxyquinoline
	Morin
	Salicylidene-o-aminophenol[a]
Au	Butylrhodamine B
	p-Dimethylaminobenzylidenerhodanine
	Rhodamine B
B	Butylrhodamine B[a]
	Benzoin[a]
	1,4-Dihydroxyanthraquinone
	Rhodamine B
	Rhodamine 6 G
Be	1-Amino-4-hydroxyanthraquinone
	3-Hydroxy-2-naphthoic acid[a]
	2-(o-Hydroxyphenyl)-benzothiazole
	8-Hydroxyquinaldine
	Morin[a]
Ca	Fluorexone[a]
	8-Quinolylhydrazone of 8-hydroxyquinaldehyde
Cd	8-(Phenylsulfonylamino)quinoline[a]
Cu	Lumocupferron[a]
Ga	5,7-Dibromo-8-hydroxyquinoline
	8-Hydroxyquinaldine
	8-Hydroxyquinoline
	Lumogallion[a]
	Rhodamine B[a]
	Rhodamine 6 G
	Sulfonaphtholazoresorcin
Ge	Rezarson[a]
Hf	Flavanol
	Morin[a]
	Quercetin
Hg	Butylrhodamine B
	Rhodamine B
In	8-Hydroxyquinaldine
	8-Hydroxyquinoline
	Rhodamine B[a]
	Rhodamine 6 G[a]
Li	8-Hydroxyquinoline
Mg	N,N′-Bis(salicylidenethylenediamine)[a]
	Lumomagneson[a]
Mo	Carminic acid

Table 4. Most widely used fluorimetric reagents in inorganic analysis

Element	Reagent
1	2
Nb	Lumogallion
Re	Butylrhodamine B 1,10-Phenanthroline Rhodamine 6 G
Sb	Rhodamine 3 B
Sc	Salicyl aldehyde, semicarbazone
Se	3,3′-Diaminobenzidine[a] 2,3-Diaminonaphthalene[a]
Sn	Flavanol 8-Hydroxyquinoline Morin
Ta	Butylrhodamine B Rhodamine B Rhodamine 6 G[a]
Te	Butylrhodamine B[a] Rhodamine 6 G
Th	1-Amino-4-hydroxyanthraquinone Morin Quercetin
Tl	Butylrhodamine B Rhodamine B[a] Rhodamine 6 G[a]
U	Rhodamine B
W	Carminic acid Flavanol
Y	5,7-Dibromo-8-hydroxyquinoline 8-Hydroxyquinoline
Zn	8-Quinolinethyol 8-(p-Tosylsulfonamido)quinoline
Zr	Flavanol Morin Quercetin

[a] These reagents are included in the rational assortment of organic reagents, recommended for fluorimetric determinations of inorganic ions by the All-Union Research Institute of Chemical Reagents and Pure Chemicals (IREA, Moscow, USSR)

4. Phosophorimetric Determinations of Elements

Phosphorescence of metal ion complexes with organic ligands is used in inorganic analysis much less frequently than fluorescence even though phosphorimetric-analytical methods are promising because the use of phosphoroscopes results in significantly lower detection limits, due to the reduction of blank magnitude caused by fluorescence of the reagent. Phosphorimetric and extraction-phosphorimetric methods have been proposed for the determination of Nb(V) by 8-hydroxyquinoline[290], Cu(II) by etioporphyrine[291,292], Be(II)[293–295], B(III)[296] by dibenzoylmethane, and Be(II) by 2-(2′-hydroxyphenyl)benzoxazole[297], Gd(III) by methyl-bis(8-hydroxy-2quinolyl)amine[298]. Recently, the phosphorimetric determination of Gd(III) by dibenzoylmethane and pyridine on paper filter at room temperature has been performed[299,300]. These methods yield low values of C_{min} (below n $\times$ 10^{-3} μg/ml) and are frequently characterized by a high selectivity. The theory of phosphorescence has not been well developed for metal ion complexes with organic ligands with the exception of dibenzoylmethane[301,302] and dimethyl ether of mesoporphyrine-IX[303–307]. The available data on the intensity of phosphorescence, relations $\varphi_{fl}/\varphi_{ph}$, τ_{ph}, values of S_1–T_1-splittings, and phosphorescence polarization as a function of the nature of the metal ion are difficult to interpret because in the complexes not only the heavy atom effect but also the effect of the magnetic field of paramagnetic ions are difficult to explain. The CI SC LCAO method has revealed that complexing of Be(II), Mg(II), Ca(II), and Sr(II) with dibenzoylmethane results in $\pi\pi^*$-phosphorescence whose maximum is shifted to the short wavelength region, in contrast to the reagent phosphorescence maximum, because of inversion of the $S_{n\pi^*}$-, $T_{n\pi^*}$-, and $T_{\pi\pi^*}$-levels in the organic moiety of the complex[308].

At present, it cannot be said which reagents are best suited for phosphorimetric determinations. In some cases, reagents with the same nature of lower excited states may be used effectively because, as shown above, they provide low blank values.

5. *Determination of Elements by Sensitized Luminescence*

The relationship between sensitized luminescence and the electronic structure of the ligand and the metal ion such as Sm(III), Eu(III), Tb(III), and Dy(III) has received much attention in the literature. In the above scheme (Fig. 4C) the quantum yields of sensitized luminescence depend on two main factors, the intramolecular energy transfer efficiency and the probability of radiationless processes[309,310]. The transfer probability is proportional to the integral of overlap of ligand phosphorescence and complexing metal ion absorption spectra[311]. Intramolecular energy transfer is most effective in complexes whose organic ligands have various low excited states, $S_{n\pi^*}$ and $T_{\pi\pi^*}$ or $S_{\pi\pi^*}$ and $T_{n\pi^*}$[312]. The radiationless processes are caused by stretching vibrations in molecules of organic ligands. The effectiveness of these processes increases as the number of vibrational quanta necessary for the overlap of the energy zone between the levels M_1 and M_0 (Fig. 4C) decreases. This explains the increased quantum yields of sensitized luminescence of lanthanide(III) complexes which contain deuterated or fluorinated molecules of organic ligands and solvents[313–323]. The probability of radiationless processes is also influenced by temperature. In some cases, its decrease leads to an increased intensity of sensitized luminescence[317,324–327]. The effect of other factors on the intensity of sensitized luminescence will be considered below.

Consequently, for determinations using sensitized luminescence, organic reagents with different lower excited states are useful. Reagents whose lower states $S_{\pi l\pi^*}$ and $T_{\pi l\pi^*}$ are characterized by high values of intersystem crossing constants, are promising for this purpose. Blanks are reduced by use of reagents whose phosphorescence quantum yield is low, for instance, those of spectral-luminescence group II.

Sensitized luminescence in inorganic analysis will be discussed below in the section on lanthanides. Fluorescence, phosphorescence and sensitized luminescence processes are independent of the electronic structure of the organic reagent and the metal ion alone. Of importance are the composition of the complex, the nature, strength, and spatial orientation of metal-ligand bonds, and conditions under which the luminescence reaction proceeds (such as pH and the nature of solvent). All these factors significantly influence the detection limit, sensitivity and selectivity of determination.

B. Determination of Elements by Their Native Luminescence

The emissions, which are characteristic of the complexing metal ion or activator itself are often referred to as native luminescence. This is true of compounds of lanthanides(III)[328–332], U(VI)[333–335], mercury-like ions: Tl(I), Sn(II), Pb(II), As(III), Sb(III), Bi(III), Se(IV), Te(IV)[21, 24, 336], transition metals: Cr(III)[337, 338], Re(I)[339], Ru(II)[340–347], Os(II)[342, 348], Pt(II)[349], Rh(III)[350–356], Ir(III)[342, 343, 348, 350, 352, 357–360], and many others[337].

The nature of their luminescence is discussed in several books[309, 361–365] and in a review[337].

Analytical chemistry makes use chiefly of the native luminescence lanthanides(III), U(VI), mercury-like ions and Cr(III) compounds. They are distinguished by temperature-quenching of luminescence. Therefore, the determination (espcially of Cr(III) and mercury-like ions) is performed at low temperatures (usually 77 K). A very simple sample cell compartment unit (Fig. 5) permits the luminescence of small analyte volumes to be measured[366]. Also, solvents which do not solidify to clear glasses at low temperature can be used. This is of practical importance.

1. Lanthanides

Lanthanides, Ln(III), are capable of displaying luminescence in solutions of inorganic salts, crystallophosphors, and complexes with organic ligands. Luminescence of inorganic salts of Ln(III) is characteristic of elements from Ce to Dy. Luminescence spectra of Ce(III), Pr(III), and Nd(III) salts are broad diffusion bands associated with transitions of electrons from the 5d- to the 4f-shell. In contrast, Sm(III), Eu(III), Tb(III), and

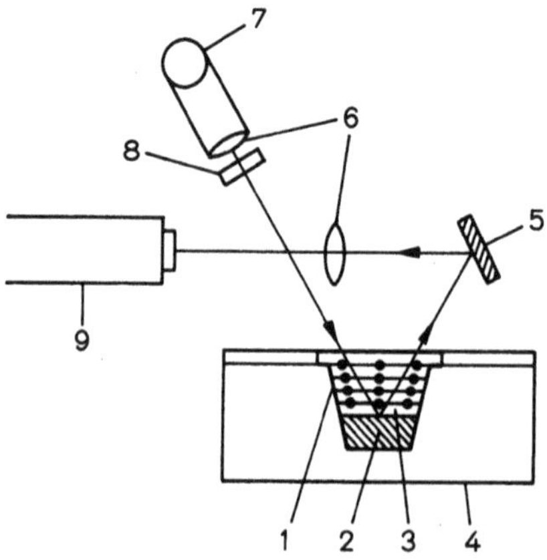

Fig. 5. Sample cell compartment unit for cryogenic luminescence measurements of frozen solutions. (1) metallic sample cell; (2) analyte solution; (3) liquid nitrogen layer; (4) sample cell compartment; (5) mirror; (6) lenses; (7) excitation source; (8) filter; (9) monochromator

Dy(III) salts are characterized by line luminescence caused by transitions of 4 f-electrons. In crystallophosphors, line luminescence of ions is caused by transitions from the partially filled 4f-shell of the lanthanides from Ce(III) to Yb(III). The crystallophosphor luminescence spectra of Ce(III) are in the far infrared, those of Gd(III) in the ultraviolet and those of Ln(III) in the visual and near infrared regions. Complexes of Sm(III), Eu(III), Tb(III), Dy(III), and Tm(III) with organic ligands exhibit line sensitized luminescence (Fig. 4C).

Luminescence of inorganic salts of Ln(III) in aqueous solutions and organic solvents containing CH, OH, NH and other groups with high-frequency vibrations is strongly quenched by conversion of excitation energy into vibrational energy due to the effect of groups[309]. This phenomenon has not yet been utilized in analytical chemistry. In aprotic solvents ($POCl_3$–$SnCl_4$ and $POCl_3$–$ZrCl_4$) the intensity of their luminescence sharply increases[367–370]. However, these solvents have hitherto not been employed in luminescence analysis of Ln(III).

Methods using luminescence of crystallophosphors activated by Ln(III) are widely applied in analysis[371–415]. These methods are systematized in a book[8] and described in[416–425]. The basis of crystallophosphors are various inorganic compounds. Luminescence of such crystallophosphors is excited in absorption bands associated with:

1) allowed 4f- 5d-transitions (Ce(III) and Tb(III));
2) forbidden transitions inside the 4f-shell (Nd(III), Dy(III), Ho(III), Er(III) and Tm(III));
3) charge transfer from O^{2-}-groups of the basis to an Ln(III) ion;
4) charge transfer from VO_4^{3-}, NbO_4^{3-}, and MoO_4^{3-}-groups of the matrix to an Ln(III) ion.

These crystallophosphors are most often photoexcited by UV light of mercury and xenon lamps, sparks, recently lasers, and also cathode and X-rays. The value of C_{min} in photoexcitation is usually 10^{-4}–10^{-6}% and in cathode or X-ray excitation, 10^{-7}–10^{-8}%. The value of C_{min} is increased 10–10^3 fold in the determination of Ln(III) in the presence of elements which quench luminescence. Studies performed during the last two years[420–425] have used dye lasers as monochromatic source for excitation crystallophosphors of Ln(III). A spectral system has been described which permits selective excitation of Ln(III) ion luminescence in absorption bands associated with transitions inside the 4f-shell[421, 423, 424]. The dependence of Ln(III) ion absorption spectra on the immediate invironment of the ion has led to methods for the determination of nonluminescent ions by emission of Ln(III) crystallophosphors such as of PO_4^{3-} (up to 1×10^{-3} μg/ml) by luminescence of $BaSO_4 \cdot Eu$[422], Y(III), La(III), Gd(III), and others of Ln(III) (up to 1×10^{-6}%) by luminescence of $CaF_2 \cdot Er$[425].

In the determination of Sm(III), Eu(III), Tb(III), and Dy(III) some methods are based on sensitized luminescence of complexes of these ions with organic ligands[327, 331, 426–476] (Table 5). The effect of electronic structure of the ligand and the metal ion on sensitized luminescence has been discussed above. The quantum yield of sensitized luminescence is in this case also influenced by the nature, strength, and direction of Ln-ligand bonds[314–316, 477–479], by the solvent[314, 325], and by the presence of impurities[480–483].

Methods of determining these metals by sensitized luminescence are highly sensitive, selective and very precise. The value of C_{min} of Eu(III) and Tb(III) is usually two orders of magnitude below those for Sm(III) and Dy(III) and may be as low as $n \times 10^{-5}$ μg/ml.

Table 5. Determination of Sm(III), Eu(III), Tb(III), Dy(III) by sensitized luminescence of their complexes with organic reagents

Reagent	Method of estimation[a]	Reference
1	2	3
Samarium(III)		
Ethylenediaminesalicyl aldehyde	Prec	426
1,10-Phenanthroline	Prec	426
2-Thenoyltrifluoroacetone and collidine	Susp	427, 428
	Extr	427, 428
2-Thenoyltrifluoroacetone and 1,3-diphenylguanidine	Susp	427, 428
	Extr	427, 428
2-Thenoyltrifluoroacetone and 1,10-phenanthroline	Susp	427, 429, 430
	Sol	427
	Extr	430–432
1,1,1,5,5,5-Hexafluoro-2,4-pentanedione	Sol	433
Hexafluoroacetylacetone and trioctylphosphine oxide	Extr	434
2-Naphthoyltrifluoroacetone and trioctylphosphine oxide	Extr	435
4-Benzoyl-3-methyl-1-phenyl-5-pyrazolone and phosphine oxides	Extr[b]	436
4-Benzoyl-3-methyl-1-phenyl-5-pyrazolone and phosphoric acid, triisobutyl ester	Extr[b]	437
Europium(III)		
Benzoylacetone	Prec[b]	331
Dibenzoylmethane	Prec	438
	Prec[b]	331, 438
	Sol	438
	Sol[b]	438, 439
5-Nitrosalicyl aldehyde	Prec	426
1,10-Phenanthroline	Prec	331, 426
	Prec[b]	331
2-Thenoyltrifluoroacetone	Prec[b]	331
	Sol	440
Dibenzoylmethane and diethanolamine	Susp	441
1,10-Phenanthroline and atophan	Susp	442, 443
1,10-Phenanthroline and novatophan	Susp	444
1,10-Phenanthroline and salicylic acid	Susp	445
	Extr	446
2-Thenoyltrifluoroacetone and collidine	Susp	427, 428
	Extr	428
2-Thenoyltrifluoroacetone and 1,3-diphenylguanidine	Susp	427, 428
	Extr	428
2-Thenoyltrifluoroacetone and 1,10-phenanthroline	Susp	427, 429
	Sol	427
	Extr	430–432, 447–449
1,1,1,5,5,5-Hexafluoro-2,4-pentanedione	Sol	450
Methyl-bis(8-hydroxy-2-quinolyl)amine	Sol	451
2-Phenylquinoline-4-carboxylic acid	Sol	452
2-Pyridinecarboxylic acid	Sol	452
2-Quinolinecarboxylic acid	Sol	452
2-Thenoyltrifluoroacetone and 2-hydroxyethylethylendiaminetriacetic acid	Sol	453, 454

Table 5 (continued)

Reagent	Method of estimation[a]	Reference
1	2	3
Benzoyltrifluoroacetone and trioctylphosphine oxide	Extr	455
Hexafluoroacetylacetone and trioctylphosphine oxide	Extr	434
1-Naphthoic acid and 1,10-phenanthroline	Extr	456
2-Naphthoyltrifluoroacetone and trioctylphosphine oxide	Extr	435
4-Benzoyl-3-methyl-1-phenyl-5-pyrazolone and phosphine oxides	Extr[b]	436, 457
4-Benzoyl-3-methyl-1-phenyl-5-pyrazolone and phosphoric acid, triisobutyl ester	Extr[b]	437, 457
4-Benzoyl-3-methyl-1-phenyl-5-pyrazolone and bases containing oxygen and nitrogen	Extr[b]	457
Terbium (III)		
Dibenzoylmethane	Prec	438
	Prec[b]	438
	Sol	438
	Sol[b]	438, 439
1,10-Phenanthroline	Prec	426
Salicylic acid, phenyl ester	Prec	426
	Susp	458
	Susp[b]	327
Antipyrine and tetraphenyl borate	Susp	459
3-Methyl-1-phenyl-5-pyrazolone and tetraphenyl borate	Susp	460
	Susp[b]	327
3-Methyl-1-tolyl-5-pyrazolone and tetraphenyl borate	Susp	460
	Susp[b]	327
Acetylacetone and EDTA	Sol	461
Bis(3-methyl-1(2)-pyridyl-5-pyrazon-1-yl)-4,4′-methane	Sol	462
2,3-Dihydroxynaphthalene and EDTA	Sol	463
α,α'-Ethylenediiminodi(o-hydroxyphenylacetic acid)	Sol	464
1,1,1,5,5,5-Hexafluoro-2,4-pentanedione	Sol	450
o-Hydroxyphenyliminodiacetic acid	Sol	465
Pyrogallolsulfonic acid	Sol	466
Salicylic acid and EDTA	Sol	467
3-Methyl-4-sulfonylphenyl-5-pyrazolone	Sol	432, 468
	Sol[b]	327
Sulfosalicylic acid and EDTA	Sol	469
	Sol[b]	327
2-Thenoyltrifluoroacetone	Sol	440
Tyrone	Sol	470
Tyrone and EDTA	Sol	471, 472
	Sol[b]	327
Tyrone and iminodiacetic acid	Sol	473
	Sol[b]	327
Hexafluoroacetylacetone and trioctylphosphine oxide	Extr	434
4-Benzoyl-3-methyl-1-phenyl-5-pyrazolone and phosphine oxides	Extr[b]	436
4-Benzoyl-3-methyl-1-phenyl-5-pyrazolone and phosphoric acid, triisobutyl ester	Extr[b]	437
Salicylic acid and antipirine	Extr	474
	Extr[b]	327

Table 5 (continued)

Reagent	Method of estimation[a]	Reference
1	2	3
Salicylic acid and 1,10-phenanthroline	Extr	446
	Extr[b]	327
Dysprosium(III)		
1,10-Phenanthroline	Prec	426
Bis-(1-phenylmethyl)-5-pyrazolone	Susp	475
3-Methyl-1-phenyl-5-pyrazolone	Susp	476
	Susp[b]	327
3-Methyl-1-tolyl-5-pyrazolone	Susp	476
	Susp[b]	327
Bis(3-methyl-1(2)-pyridyl-5-pyrazon-1-yl)-4,4′-methane	Sol	462
3-Methyl-4-sulfonylphenyl-5-pyrazolone	Sol	432, 468
	Sol[b]	327
Tyrone and EDTA	Sol	471, 472
	Sol[b]	327
Tyrone and iminodiacetic acid	Sol	473
	Sol[b]	327
Salicylic acid and 1,10-phenanthroline	Extr[b]	327
4-Benzoyl-3-methyl-1-phenyl-5-pyrazolone and phosphine oxides	Extr[b]	436

[a] Luminescence of precipitates (Prec), suspensions (Susp), solutions (Sol), extracts (Extr) of metal ion complexes

[b] Determination at 77 K

According to the determination technique, luminescence is classified into that of precipitates, suspensions, solutions or extracts of metal ion complexes (Table 5). Methods which make use of the luminescence of precipitates or suspensions of complexes are most sensitive ($C_{min} \sim 10^{-4} - 10^{-5}$ μg/ml). This is attributed to the high concentration of the radiating centers per unit volume and low effectiveness of radiationless deactivation of excitation energy by high frequency vibrations of bonds in ligand and solvent molecules. In this case, however, the complexes decompose in UV light. Furthermore, in the determination of Ln(III) as precipitates or suspensions of complexes, the luminescence intensity strongly depends on the presence of impurities, which results in reduced selectivity. The selectivity of the determination of Ln(III) increases, especially in the presence of luminescence quenching ions, if solutions of complexes are used in the analysis. However, C_{min} in this case increases by more than one order of magnitude compared with the determination of Ln(III) as precipitates or suspensions of complexes. Of all the methods described here, the most selective are extraction-luminescence methods whereby C_{min} is as low as $n \times 10^{-5}$ μg/ml. In some cases, the detection limit can be enhanced by decreasing the temperature[327]. However, the application of low termperatures to the determination of Ln(III) has not yet been sufficiently utilized.

2. *Mercury-like Ions*

Mercury-like ions Tl(I), Sn(II), Pb(II), As(III), Sb(III), Bi(III), Se(IV), and Te(IV) have the electron shell of a mercury atom, $1s^2 \ldots np^6nd^{10}(n+1)s^2$. They exhibit phosphorescence in alkali halides, oxides and other crystallophosphors in frozen aqueous solutions and extracts of their halide complexes. In analytical chemistry, phosphorescence of halide complexes of mercury-like ions at low (77 K) temperatures[336, 484–500] is most widely used (Table 6). Halide complexes of Tl(I), Sn(II), and Pb(II) display phos-

Table 6[a]. Determination of mercury-like ions by low-temperature (77 K) luminescence

Ion	Medium	λ(nm) or filter excit.	emiss.	Detection limit (μg/ml)	Ref.	Interferences
1	2	3	4	5	6	7
Tl(I)	HCl	242	396	5×10^{-2}	484	2-Fold excess of Pb(II), 70-fold excess of Bi(III) and 3-fold excess of Te(IV) did not interfere with excitation by monochromatic light with wavelength 254 nm[486, 487]
		254	370	2×10^{-2}	485, 486	
		256	380	2×10^{-1}	488	
		Scan.	396	1×10^{-2}	484	
	HBr	223, 263	428	5×10^{-2}	484	
		265	400	2×10^{-2}	485	
		Scan.	428	2×10^{-2}	484	
Sn(II)	HCl	220, 268	585	1	489	
	HBr	265, 304	574, 612	1	489	
Pb(II)	HCl	260–270	385	1×10^{-2}	490	300-Fold excess of Tl(I), 1000-fold excess of Bi(III) and 50-fold excess of Te(IV) did not interfere with excitation by monochromatic light with wavelength 270 nm[486, 487] 10-Fold excess of Te(IV) did not interfere with determination of 0,3 μg/ml Pb(II), 15-fold excess of Bi(III) and 2000-fold excess of Sn(IV)-1 μg/ml Pb(II), 300-fold excess of Sn(IV) – 10 μg/ml Pb(II), 60-fold excess of Sn(IV) – 50 μg/ml Pb(II)[491, 492]
		260–270	Visual.	1	490	
		270	385	8×10^{-3}	486, 487	
		272	423	1×10^{-2}	484	
		276	390	2×10^{-3}	488	
		Scan.	423	1×10^{-3}	484	
		272[b]	420	5×10^{-3}	493	
		UFS-2[b]	420	5×10^{-2}	493	
		UFS-1[b]	Visual.	1	494	
		UFS-2[b]	Visual.	2	494	
		UFS-2[c]	484	1×10^{-2}	495	
	HBr	302	424	1×10^{-2}	484	1000-Fold excess of alkali, alkaline earth metals, Cd(II), Sn(II), Be(II), Co(II), Ni(II), Cr(III), In(III), Al(III), Sb(III), Bi(III), Se(IV), Te(IV), and As(V) did
		313	420	1×10^{-2}	336	
		Scan.	424	2×10^{-3}	484	
		UFS-2	Visual.	1×10^{-1}	336	

Table 6[a] (continued)

Ion	Medium	λ(nm) or filter excit.	emiss.	Detection limit (μg/ml)	Ref.	Interferences
1	2	3	4	5	6	7
						not interfere with the determination of Pb(II) by increase of luminescence intensity (from 77 to 293 K)[29, 493]
As(III)	HCl	312	475	2×10^{-1}	496	
	HBr	350	690	8×10^{-3}	496	
		Scan.	690	8×10^{-3}	496	
Sb(III)	HCl	260, 314	625	1×10^{-1}	497	100-Fold excess of Tl(I), Ag(I), Pb(II), Ca(II), Mg(II), Ba(II), Sr(II), Be(II), Cd(II), Mn(II), Ni(II), Al(III), Cr(III), Te(IV), 10-fold excess of Cu(II), Fe(III) did not interfere[487]
		306	582	2×10^{-1}	488	
		UFS-2	620	2×10^{-1}	336	
		UFS-2	Visual.	5×10^{-1}	336	
	HBr	360	640	1×10^{-3}	497	2000-Fold excess of Pb(II), 1500-fold excess Bi(III) did not interfere with the determination of 1 μg/ml Sb(III); 10-fold excess of Pb(II) – 50 μg/ml Sb(III)[491]
		366	586	1×10^{-2}	498	
		366	640	3×10^{-3}	486, 487	
		UFS-1	Visual.	1×10^{-1}	494	
		UFS-2	Visual.	1×10^{-1}	494	
		UFS-3	Visual.	1×10^{-1}	494	
		UFS-3	Visual.	1×10^{-2}	487	
Bi(III)	HCl	313	410	2×10^{-3}	486, 487	500-Fold excess of Tl(I), 2000-fold excess of Pb(II), 70-fold excess of Te(IV) did not interfere with the excitation of monochromatic light of 313 nm[487]
		313	410	1×10^{-2}	490	
		313	Visual.	5×10^{-2}	490	
		330	410	2×10^{-3}	488	
		340	420	2×10^{-3}	497	
		Scan.	420	1×10^{-3}	497	
		UFS-2	410	8×10^{-3}	336	Equal amounts of Te(IV) did not interfere with the determination of 1 μg/ml Bi(III) (filter UFS-2); 10-fold excess of Te(IV) at λ_{ex} = 313 nm.
		UFS-1	Visual.	1	494	
		UFS-2	Visual.	1	494	
		UFS-3	Visual.	1	494	500-Fold excess of Pb(II), 300-fold excess of Te(IV), 200-fold excess of Sn(IV) did not interfere with the determination of 0,1 μg/ml Bi(III) at λ_{ex} = 313 nm[491]
	Extraction with TBP	313	480	4×10^{-2}	499	30-Fold excess of Te(IV) did not interfere with extraction-phosphorimetric

Table 6[a] (continued)

Ion	Medium	λ(nm) or filter excit.	emiss.	Detection limit (μg/ml)	Ref.	Interferences
1	2	3	4	5	6	7
	Extraction with butyl alcohol	313	460	1×10^{-1}	499	determination of 30 μg Bi(III) in the case of extraction with butyl alcohol from 0,5 M HCl[499]
	HBr	380	487	1×10^{-1}	497	
		Scan.	487	1×10^{-2}	497	
	Extraction with TBP	313	510	7×10^{-2}	499	
Se(IV)	HCl	305, 330	390	4×10^{-1}	497	
	HBr	352	550	8×10^{-1}	497	
Te(IV)	HCl	262, 313, 375	705	1×10^{-3}	489	50-Fold excess of Tl(I), 20-fold excess of Pb(II), 10-fold excess of Bi(III) did not interfere[487]
		313, 365	640	4×10^{-2}	487	
		380	586	2×10^{-2}	500	Determination of up to 3×10^{-2}% Te in Pb compounds is possible with excitation by monochromatic light at 380 nm[500]
	Extraction with octyl alcohol	UFS-3	660	3×10^{-2}	499	3000-Fold excess of Pb(II), 1000-fold excess of Bi(III), 2000-fold excess of Sn(IV) did not interfere with the determination of 1 μg/ml Te (IV);
	HBr	UFS-3	695	1×10^{-2}	499	1000-fold excess of Pb(II), 200-fold excess Sn(IV) did not interfere with the determination of 10 μg/ml Te(IV); 500-fold excess of Pb(II), 40-fold excess of SN(IV) – 50 μg/ml Te(IV)[492]
	Extraction with TBP	UFS-3	690	1×10^{-3}	499	
	Extraction with hexyl alcohol	UFS-3	695	5×10^{-3}	499	50-Fold excess of Sb(III) did not interfere with extraction-phosphorimetric determination of 30 μg Te(IV) in the case of extraction with octyl alcohol from 11 M HCl
	Extraction with ethyl-acetate	UFS-3	695	5×10^{-3}	499	

[a] Some data used from Ref. 336
[b] Determination by increase of luminescence intensity from 77 to 293 K
[c] Determination at 203 K

phorescence weakly at room temperature and strongly at low temperatures. Phosphorescence of solutions of halide complexes of As(III), Sb(III), Bi(III), Se(IV), and Te(IV) is observed only upon cooling. Phosphorescence studies of halide complexes of Tl(I)[484–488, 501–511], Sn(II)[489, 505–508, 510, 512–515], Pb(II)[484, 486–488, 490–493, 495, 499, 502–511, 513, 516–519], As(III)[496, 508, 510, 513, 520], Sb(III)[486–488, 491, 492, 497–499, 508, 510, 513, 514, 521], Bi(III)[486–488, 490–492, 497, 499, 508–510, 514, 519], Se(IV)[497, 513, 520, 522], Te(IV)[487, 489, 491, 492, 499, 500, 508, 513, 523, 525], have shown that the nature of emission or radiation is largely similar to that of their alkali halide crystallophosphors[526–542]. The absorption and phosphorescence center in both cases is a mercury-like ion whose energy levels are deformed upon interactions with the environment. Under the influence of the environment, the energy levels of ions in halide complexes come closer together than in crystals, which results in an insignificant bathochromic shift of absorption spectra and, to a larger extent, phosphorescence. The ground state of the mercury-like ion is associated with the term $^1S_0$ and the first excited state, $ls^2 \ldots np^6nd^{10}$ (n+1)s(n+1)p, with the terms $^3P_0$, $^3P_1$, $^3P_2$, and $^1P_1$. The absorption spectra consist of one broad band ($^1S_0 \rightarrow {}^1P_1$-transition) in the short wavelength and three poorly resolved bands ($^1S_0 \rightarrow {}^3P_{0,1,2}$) in the long wavelength regions. Phosphorescence spectra include overlapping bands which are associated with $^3P_{0,1,2} \rightarrow {}^1S_0$. In terms of selection rules the most probable are $^3P_0 \rightarrow {}^1S_0$- and $^3P_1 \rightarrow {}^1S_0$-transitions. Emission of radiation of halide complexes of mercury-like ions is excited only in the absorption region caused by $^1S_0 \rightarrow {}^3P_{0,1,2}$-transitions. Since the energy of transitions differs from ion to ion. (This is demonstrated in Fig. 6 by the position of the energy levels of A – free mercury-like ions and B – Tl(I), Sn(II), and Pb(II) in KCl-based crystallophosphors) their excitation and phosphorescence spectra are also different. In the excitation in the 248–253 nm range phosphorescence occurs only in chloride complexes of Tl(I), in the range 265–270 nm – in chloride complexes of Pb(II) and Te(IV), at 313 nm – in chloride complexes of Bi(III) and Te(IV)[486, 487, 490]; the phosphorescence spectrum of the Te(IV) complex is shifted to wavelengths longer than those of Pb(II) and Bi(III) complexes. By changing the conditions for excitation and registration of phosphorescence of frozen solutions of halide complexes, mercury-like ions are determined without preliminary separation[486, 487, 490–492]. The value of C_{min} largely depends on the excitation conditions (Table 6). The low temperature phosphorimetric methods of determining mercury-like ions were developed in the analysis of highly pure substances[21, 24, 336, 543] and semiconducting materials[491, 492].

In the rise of the temperature of frozen Tl(I), Pb(II) and Bi(III) halide complexes solutions, quantum yields increase stepwise and a bathochromic shift of radiation spectra is observed which is usually attributed to a change in the matrix structure with is usually attributed to a change in the matrix structure with increasing temperature[505–511]. This phenomenon is utilized in a highly selective and sensitive method of determining Pb(II)[292, 493] in HCl (1×10^{-7}%), NaH_2PO_4 (1×10^{-5}%), ZnS (3×10^{-5}%), and C_2H_5OH (1×10^{-7}%).

Luminescence is also a feature of frozen extracts of chloride and bromide complexes of Pb(II), Sb(III), Bi(III), and Te(IV)[499]. Their excitation spectra practically do not change and phosphorescence is shifted to wavelengths longer than those of frozen aqueous solutions. The bathochromic shift of phosphorescence spectra decreases in the order Pb(II) > Bi(III) > Sb(III) > Te(IV). The extracting agent also affects the spectra in that replacement of alcohols by esters and TBP leads to bathochromic shift of phorsphorescence spectra of frozen extracts. The intensity of the phosphorescence of these extracts

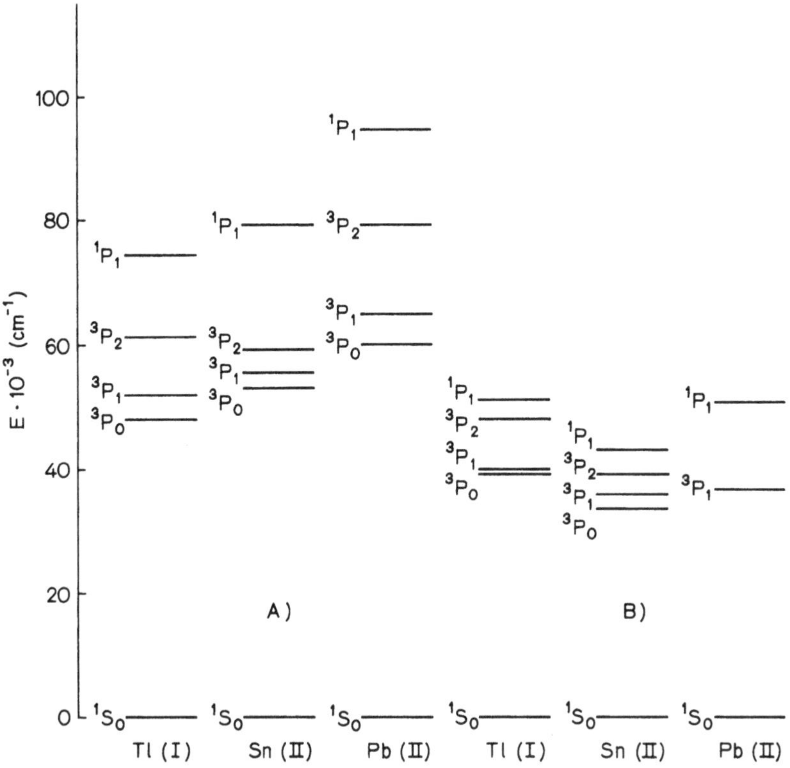

Fig. 6. Electronic levels of (A) free mercury-like ions of Tl(I), Sn(II), Pb(II), (B) the same ions but in crystallophosphors on the basis of KCl

depends on the nature of the extracting agent and concentration of the halogen acid. The luminescence is most intensive in TBP extracts and alcohols. The phosphorescence is maximal in the extract of Pb(II) and Bi(III) complexes from diluted acids (0.5–2.0 M HCl or HBr) and of Sb(III) and Te(IV) complexes from 7.0–11.0 M HCl or 3.5–5.0 M HBr. Differences in the extraction conditions for halide complexes of mercury-like ions and phosphorescence of their frozen extracts are utilized in highly sensitive and selective methods of determining Bi(III) and Te(IV)[499, 543]. Thus, up to $5 \times 10^{-9}\%$ Te(IV) in HBr and $4 \times 10^{-5}\%$ in ZnSe[543] can be determined.

3. *Chromium*

Cr(III) exhibits luminescence in crystallophosphors, in frozen solutions and in extracts of Cr(III) complexes with inorganic and organic ligands[309, 337, 544–563].

Its electronic configuration is $1s^2 \ldots 3p^6\, 3d^3$. The position of energy levels in its octahedric complexes is shown in Fig. 7. Depending on the ligand field strength in these complexes the lower excited states may be ^2E_g and $^4T_{2g}$, and $^4T_{2g} \rightarrow {}^4A_{2g}$-fluorescence and/or $^2E_g \rightarrow {}^4A_{2g}$-phosphorescence can be observed[337]. The structure of phosphorescence spectra is more regular than that of fluorescence spectra, which makes the phos-

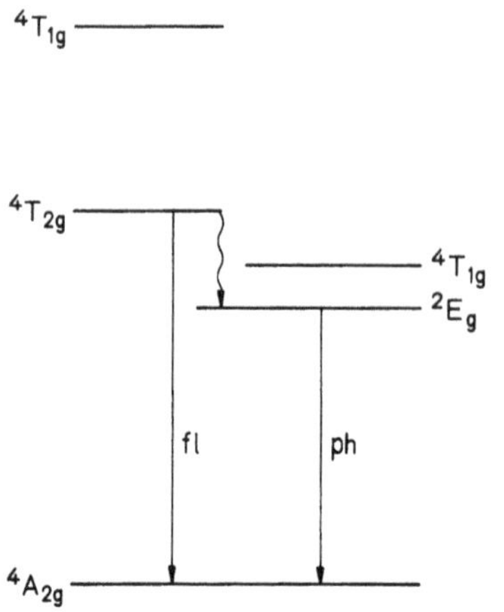

Fig. 7. Electronic levels and electronic transitions in octahedral Cr(III) complexes

phorimetric method of determining Cr(III) more valuable than the fluorimetric method. However, characteristic of all Cr(III) complexes is temperature quenching of phosphorescence[309, 337]. In particular, for Cr(III) compounds which are coordinated through oxygen atoms the temperature quenching barriers are so low that a temperature decrease to 77 K does not eliminate this quenching which only ceases at 4 K[559]. In contrast, for Cr(III) complexes where coordination is effected through nitrogen atoms, the barrier is high enough (~ 1000 cm^{-1}) and the probability of deactivating the ^2E_g-level through the $^4T_{2g}$-level is reduced at much higher temperatures[547] (for instance, for ammonia complexes at 223 K[548]). Phosphorescence of Cr(III) rhodanide complexes at 77 K is chiefly used in analytical chemistry. A highly selective phosphorimetric[564] method and extraction-phosphorimetric[366, 565] methods have been developed for the determination of Cr(III) ($C_{min} = 1 \times 10^{-4}$ μg/ml) (e.g. detection of Cr(III) in SiO_2[564] and waste water[366, 565]).

Very interesting are studies on sensitized luminescence of Cr(III) complexes with organic ligands such as 8-hydroxyquinoline, its derivatives, and β-diketones[120, 544–546, 566]. The probability of radiationless transfer of excitation energy, as in the case of Ln(III) compounds, is a function of the nature and relative position of energy levels in the ligand and metal ion. Thus, for 8-hydroxyquinolinate and 8-hydroxy-quinaldinate of Cr(III), excitation of the ligand absorption bands results in sensitized $^2E_g \rightarrow {}^4A_{2g}$-phosphorescence. On the other hand Cr(III) 5,7-dibromo-8-hydroxyquinolinate exhibits weak molecular fluorescence. Despite its promising aspects, sensitized luminescence of Cr(III) complexes with organic ligands has not yet found wide application in analytical chemistry.

4. Uranium

Many inorganic compounds of U (III, IV and VI) may exhibit luminescence. Analytical chemistry makes use of fluorescence of UO_2^{2+} complexes with inorganic and organic ligands and of U(VI) crystallophosphors.

The electronic configuration of U(VI) is $1s^2 \ldots 5s^2\,5p^6\,5d^{10}\,6s^2\,6p^6$ and that of the O^{2-} $1s^2\,2s^2\,2p^6$. Superposition of these atomic orbitals results in molecular orbitals of the UO_2^{2+}. The nature of orbitals, and particularly the contribution of unfilled 5f-, 6d- and

7 s-orbitals of U(VI) to the formation of molecular orbitals, has not yet been ascertained[334]. Ionorganic salts of U(VI) absorb intensively in the 200–300 nm region and much weaker, with a pronounced structure, in the 330–550 nm region. The nature of absorption is still in doubt. Some authors believe that short wavelength absorption is caused by transition of electrons from 5 f- or 6 d-orbitals of U(VI)[567, 568]; others attribute this absorption to transitions to levels coordinating with the UO_2^{2+} of ligands[569]. Most papers are concerned with structural absorption spectra of U(VI) salts in the 330–550 nm range[570–586]. Some authors believe[572, 574] that this absorption is caused by a single electronic state while many others assume that U(VI) salts have several excited states[570, 571, 573, 575–586]. There is, however, no agreement on their nature. Studies of absorption and luminescence spectra of perchlorate U(VI) solutions have revealed 24 transitions[582] which are grouped into seven main bands. The first two excited states of UO_2^{2+} with energies of 20502 and 21270 cm^{-1} are regarded as triplet, the next five as singlet. The same conclusion on the nature of lower electronically excited states of U(VI) salts was drawn from quantum mechanical computations[567, 568]. Many researchers believe that this conclusion is correct[577–580, 584]. Several Soviet-scientists[569–571, 575, 576, 581, 583, 585], however, have come, due to extensive experimental evidence and literature data, to the conclusion that absorptions in the 350–550 nm range represent electronic transitions to four excited singlet states of the UO_2^{2+} while absorptions in the 200–300 nm range are caused by electronic transitions involving molecular orbitals of ligands. In the corresponding papers the excited states of U(VI) are classified according to the symmetry type as well as multiplicity. In later research, flash photolysis has convincingly proved the singlet nature of radiation from U(VI) salts, and energies of higher singlet states have been computed[587]. Fluorescence spectra of U(VI) salts are observed in the 520–620 nm range and, like absorption spectra, have several bands. Fluorescence of such salts has a rather long lifetime ($\tau_{fl} \sim 10^{-4}$ s) and $\varphi_{fl} \sim 1$ in the absence of quenchers. In analytical chemistry, fluorescence of U(VI) is utilized in aqueous solutions of $Na_3P_3O_9$, HF, H_3PO_4, and H_2SO_4. Fluorescence intensity depends on the nature of the ligand and its concentration, acidity of the medium, ionic state of U(VI), and the presence of impurities. To separate U(VI) ions from most elements, they are usually extracted by TBP solutions in the presence of EDTA and then reextracted by $Na_3P_3O_9$ solution or the associated acid and fluorimetric readings are taken (C_{min} up 1×10^{-2} μg/ml). Many organic compounds quench fluorescence of U(VI) salts in aqueous solutions because of intrmolecular energy or electron transfer[588]. Decreasing the temperature to 77 K leads, however, to intensive fluorescence. A rather selective fluorimetric method for determining U(VI) salts in frozen TBP extracts (C_{min} = 10 μg/ml) has been described[589]. A lower value of C_{min} (1×10^{-2} μg/ml) is obtained by a method which makes use of fluorescence of frozen U(VI) thenoyltrifluoracetonate[590].

The most widely applied method involves fluorimetric determination of U(VI) as crystallophosphor compounds. Their matrixes can be phosphates and carbonates of alkali and alkaline earth metals, and particularly NaF. In the latter case, as little as 1×10^{-5} μg U(VI) can be determined. Since transition metal ions are strong quenchers of U(VI) fluorescence in crystallophosphors, they are previously separated and the determination is carried out by the method of constant addition.

The extensive literature on fluorimetric determinations of U(VI) and the investigation of the fluorescence of U(VI) compounds have been reporte in a monograph[335] and in several papers[591–604].

C. *Determination of Elements by Chemiluminescence*

Chemiluminescence is the emission of radiation caused by energy which is released by a chemical reaction. It is most often observed in liquid phase reactions in the oxidation of organic and, infrequently, inorganic substances by hydrogen peroxide, molecular oxygen, hypochlorite, and other oxidants or in the gas phase[605)]. A feature common to all chemiluminescent reactions are exothermally elementary steps which result in a non-equilibrium of electronically excited states of reaction products. These states can be deactivated by radiation of light quanta (chemiluminescence). Chemiluminescence can be regarded as a two-step process involving excitation (41) and radiation (42)

A + B	$\rightarrow$ P*	+ other products	(41)
reactants	excited product		
P*	$\rightarrow$ P + hν		(42)

This review is too short for a detailed description of the various mechanisms and characteristics of chemiluminescence. These have been discussed in recent reviews[606–610)] and monographs[611, 612)]. It should, however be mentioned that chemiluminescence is often accompanied by radical chain reactions, including redox processes, proceeding by a free radical mechanism. The chemiluminescence intensity I_{CL} is equal to

$$I_{CL} = \varphi_{ex} \cdot \varphi_{fl} \cdot \vartheta \tag{43}$$

where φ_{ex} and φ_{fl} are quantum yields of excitation and fluorescence of the product P* and ϑ is the rate of reaction (41).

The luminescence quantum yield φ_{CL} is defined as

$$\varphi_{CL} = \varphi_{ex} \cdot \varphi_{fl} \tag{44}$$

Then

$$I_{CL} = \varphi_{CL} \cdot \vartheta \tag{45}$$

Chemiluminescence analysis of inorganic substances makes use of the ability of many elements with an unfilled d-shell to quench fluorescence (reduction of φ_{fl}), to catalyze, and, more rarely, to inhibit (increase or decrease of ϑ) chemiluminescence reactions, the change of I_{CL} being proportional to the concentration of the element. Chemiluminescence analysis most often uses oxidation of luminol, luceginin and, infrequently, lophin and siloxen by hydrogen peroxide in alkaline medium[606, 611, 612)]. Chemiluminescence is recorded by a fast photographic film and, more effectively by photoelectronic multipliers. Recently, various chemiluminescence analyzers with photoelectric recording, which increase the speed and improve the precision of analysis, have been suggested[613–618)]. Also, highly sensitive photomultipliers which can register φ_{fl} as low as 10^{-15} [612)] and the absence of blank in chemiluminescence reactions significantly reduce C_{min}. For most elements, $C_{min} \sim n \times 10^{-3}$ μg/ml and in many cases, $n \times 10^{-5}$ μg/ml, which makes the chemiluminescence analysis one of the most sensitive instrumental methods.

Its disadvantage is low selectivity. This is improved either by preliminary separation or by using the difference in the catalytic or kinetic activity of complexes of the elements. Thus, rather selective chemiluminescence methods have been developed for the determinations of Fe(III), Cr(III) and Co(III) which are based on the different catalytic action of the complexes with EDTA (Fe and Cr) and 2-hydroxyethylethylenediaminetriacetic acid (Co) on luminol oxidation by hydrogen peroxide[619]. The same reaction has been applied to the highly selective determination of Cr(III) by using the kinetic inertness of its complexes with EDTA[614, 615, 620–622]. Investigations into extraction-chemiluminescence determination methods have also started[623, 624].

Chemiluminescence is widely used in inorganc analysis (Fig. 3); only in recent years have chemiluminescence reactions been proposed for the determination of over 30 elements[613–615, 617–663] (Table 7).

Table 7[a]. Chemiluminescence methods of determining elements developed in recent years

Metal ion	Element ion action[b]	Reagents[c]	Registration[d]	Detection limit (μg/ml)	Reference
1	2	3	4	5	6
Ag(I)	Cat	GA–H_2O_2	PE	5×10^{-1}	617
	Cat	Lm–$K_2S_2O_8$-2,2′-dipyridyl	PG	1×10^{-2}	625
	Cat	Lm–$K_2S_2O_8$–NH_3	PG	2×10^{-1}	626
	Inh	Lm–$CuSO_4$–KCN	PG	1×10^{-3}	627
As(V)	Ox	Lm–$(NH_4)_2MoO_4$–NH_4VO_3	PG	6×10^{-3}	623
	Ox	Lm–$(NH_4)_2MoO_4$–NH_4VO_3-2-propanol	PG	1×10^{-3}	623
Au(III)	Ox	Lm–Cl^-	PG	1×10^{-2}	625, 628, 629
Bi(III)	Cat	Lc–H_2O_2	PG	2×10^{-1}	630
Cd(II)	Cat	GA–H_2O_2	PE	8×10^{-1}	617
Ce(III)	Cat	Lc–H_2O_2	PG	8×10^{-1}	631
Co(II)	Cat	GA–H_2O_2	PE	4×10^{-4}	617
	Cat	Lc–H_2O_2	PG	6×10^{-4}	632
	Cat	Lc–H_2O_2	PE	2×10^{-5}	633, 634
	Cat	Lm–H_2O_2	PE	5×10^{-1}	613
	Cat	Lm–H_2O_2–EDTA	PE	6×10^{-6}	618
	Cat	Lm–H_2O_2–EDTA-2-hydroxyethylethylenediaminetriacetic acid	PE	1×10^{-4}	619
Cr(III)	Cat	BSHCBA–KIO_4	PG	2×10^{-3}	635
	Cat	BSHCBA–KIO_4	PE	2×10^{-3}	635
	Cat	Lc–H_2O_2-methyl alcohol	PE	5×10^{-3}	636
	Cat	Lc–H_2O_2–EDTA	PG	5×10^{-2}	637
	Cat	Lm–H_2O_2–EDTA	PE	5×10^{-2}	614
	Cat	Lm–H_2O_2–EDTA	PE	2×10^{-5}	615, 620
	Cat	Lm–H_2O_2–EDTA	PE	1×10^{-5}	622
Cu(II)	Cat	Lm–H_2O_2	PE	5×10^{-1}	613
	Inh	Lc–H_2O_2–$CoSO_4$	PG	2×10^{-3}	638
	Inh	Lc–H_2O_2–$MnSO_4$	PG	3×10^{-4}	639
	Ox	Lm–KCN	PE	1×10^{-4}	640

Table 7[a] (continued)

Metal ion	Element ion action[b]	Reagents[c]	Registration[d]	Detection limit (μg/ml)	Reference
1	2	3	4	5	6
Fe(III)	Cat	Lm–H_2O_2–EDTA	PE	3×10^{-3}	619
Fe(II)	Red	Lc	PE	6×10^{-3}	641
Hf(IV)	Inh	Lm–H_2O_2–$CuSO_4$	PG	2×10^{-2}	642
	Inh	Lm–H_2O_2–$CuSO_4$	PE	1×10^{-3}	642
Hg(II)	Inh	Lm–$CuSO_4$–KCN	PG	2×10^{-3}	627
Ir(IV)	Cat	BSHCBA–KIO_4	PG	2×10^{-4}	635
	Cat	BSHCBA–KIO_4	PE	4×10^{-4}	635
	Cat	Lm–H_2O_2	PG	1×10^{-2}	643
Ir(III, IV)	Cat	Lm–KIO_4	PG	1×10^{-2}	644, 645
Ir(IV)	Ox	Lm	PG	4×10^{-2}	644, 645
Mn(II)	Cat	BSHCBA–KIO_4	PG	8×10^{-3}	635
	Cat	BSHCBA–KIO_4	PE	4×10^{-3}	635
	Cat	GA–H_2O_2	PE	4×10^{-1}	617
	Cat	Lm–H_2O_2	PE	5×10^{-1}	613
	Cat	Lm–H_2O_2-1,10-phenanthroline-citrates	PG	5×10^{-4}	646
Mo(V)	Cat	Lm–H_2O_2	PE	1	647
	Red	Lc	PE	3×10^{-2}	648
Ni(II)	Cat	Lc–H_2O_2	PG	2	649
	Cat	Lm–$K_2S_2O_8$	PG	2×10^{-2}	650
	Cat	Lm-percaprinic acid	PG	2×10^{-4}	651
	Cat	Lm-percaprinic acid	PE	1×10^{-3}	652
Os(VIII)	Cat	BSHCBA–KIO_4	PC	5×10^{-3}	635
	Cat	BSHCBA–KIO_4	PE	6×10^{-3}	635
Os(IV)	Cat	Lm–H_2O_2	PG	2×10^{-3}	625
Os(VI, VIII)	Cat	Lm–H_2O_2	PG	2×10^{-4}	625, 653
Os(III)	Cat	Lm–H_2O_2	PE	1×10^{-2}	654
Os(IV)	Cat	Lm–H_2O_2	PE	1×10^{-3}	654
Os(VI, VIII)	Cat	Lm–H_2O_2	PE	1×10^{-4}	653, 654
P(V)	Ox	Lm–$(NH_4)_2MoO_4$–NH_4VO_3	PG	3×10^{-3}	655
	Ox	Lm–$(NH_4)_2MoO_4$–NH_4VO_3-butyl alcohol	PG	7×10^{-4}	624
Pb(II)	Cat	GA–H_2O_2	PE	1	617
	Cat	Lm–H_2O_2	PE	1	613
Pt(IV)	Cat	Lm–H_2O_2–I^-, SCN^-	PG	5×10^{-1}	643
	Cat	Lm–H_2O_2	PE	1×10^{-5}	656
Rh(III)	Cat	Lm–KIO_4	PG	4	657
	Cat	Lm–KIO_4	PG	4×10^{-1}	645
	Cat	Lm–KIO_4	PG	4×10^{-2}	644
Ru(IV)	Cat	BSHCBA–KIO_4	PG, PE	1×10^{-2}	635
	Cat	Lm–H_2O_2	PG	1×10^{-3}	643
Ru(III, IV)	Cat	Lm–H_2O_2	PE	1×10^{-2}	658

Table 7[a] (continued)

Metal ion	Element ion action[b]	Reagents[c]	Registration[d]	Detection limit (μg/ml)	Reference
1	2	3	4	5	6
	Cat	Lm–H_2O_2–EDTA	PG	3×10^{-4}	625, 659
Ru(III, IV, VI, VII)	Cat	Lm–KIO_4	PG	3×10^{-3}	625, 659
Ru(IV)	Cat	Lm–$K_2S_2O_8$	PG	6×10^{-2}	659
Sb(V)	Cat	Lm–H_2O_2-dibutyl ether	PG	5×10^{-3}	660
Si(IV)	Ox	Lm–$(NH_4)_2MoO_4$	PG	1×10^{-2}	655
	Ox	Lm–$(NH_4)_2MoO_4$-butyl alcohol	PG	2×10^{-3}	624
Ti(IV)	Inh	Lm–H_2O_2–$CuSO_4$	PG	1×10^{-2}	642
	Inh	Lm–H_2O_2–$CuSO_4$	PE	5×10^{-4}	642
V(IV)	Cat	Lm–O_2	PG	1×10^{-2}	661
	Cat	Lm–O_2–$P_2O_7^{4-}$	PG	2×10^{-3}	662
	Cat	Lm–H_2O_2	PG	1×10^{-2}	663
	Cat	Lm–H_2O_2	PG	1×10^{-3}	661, 662
	Cat	Lm–KIO_4	PG	1×10^{-2}	663
	Ox	Lm–$(NH_4)_2MoO_4$–$(NH_4)_3PO_4$	PG	5×10^{-3}	663
	Red	Lc	PG	5×10^{-3}	648

[a] Some data used from Ref. 1
[b] Cat = catalyst, Inh = inhibitor; Ox = oxydant; Red = reductant
[c] BSHCBA = p-chlorobenzoic acid, 5-bromosalicylidenehydrazide, GA = gallic acid; Lc = lucigenin; Lm = luminol
[d] PG = photographic; PE = photoelectric

D. Luminescence Determination of Non-metals

Relatively few reliable luminescence reactions for the determination of non-metals are known. Although the classification of elements into metals and non-metals is rather conventional, the determination of the latter has some outstanding features. In principle the luminescence of any organic compound can be used for the determination of one of its elements. Thus, nitrogen can be determined when converted into a fluorescent aromatic amine. Such reactions are, however, rather complicated and are not used in analysis. Therefore, indirect methods of fluorimetric analysis are applied to the determination of nonmetals. These methods are classified into two types. In the one, a nonfluorescent ion reacts with a fluorescent compound to form nonfluorescent products (these methods are sometimes classified as quenching methods). One of such ions for which no direct luminescence methods have been developed is F^-. In most cases, F^- are determined from the decrease in intensity of fluorescing Al(III) complexes with 8-hydroxyquinoline or quercetin[4, 6] which results in the formation of a more stable and nonfluorescent AlF_6^{3-} complex. Recently, complexes of Al(III) with 1-(2-pyridylazo)-2-naphthole[664], Th(IV) with morin[665], and Zr(IV) with calcein blue[666] and flavanol[667] have been proposed. In indirect methods of the other type, the ion reacts with the nonfluorescent compound; the product is then either fluorescent itself or can be converted into a new fluorescent compound. This can be illustrated by the indirect determination of CN^- [668]:

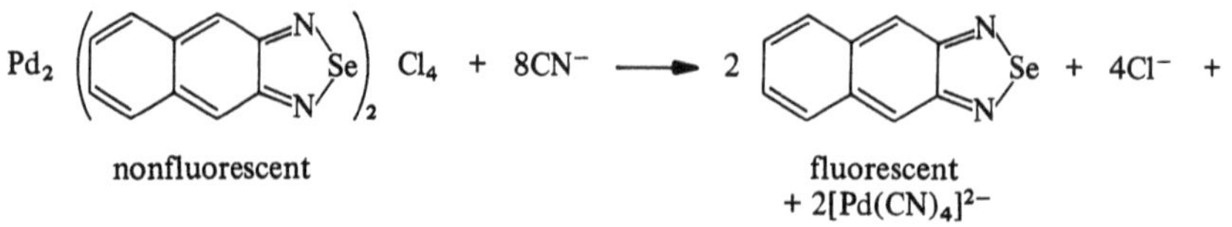

In this reaction, a nonfluorescent complex of Pd(II) is transformed into intensively fluorescent 4,5-benzopiazselenole. Indirect fluorimetric methods are used for the deterination of the SO_4^{2-} (complexes of Th(IV) with morin[669] and salicylfluorone[670], Zr(IV) with calcein blue[671]), S^{2-} (complexes of Cu(II) with 2-(o-hydroxyphenyl)benzoxazole[672] and Hg(II) with 2,2′-pyridylbenzimidazole[673]). Such methods are rather selective but not very sensitive. For this reason, direct luminescence methods are of special interest. Some of them, e.g. fluorimetric and phosphorimetric determination of B(III); As(III) and Se(IV) by their native luminescence and of As(V), P(V) and Si(IV) by chemiluminescence have been discussed above (Tables 3, 6, 7). The most characteristic methods of determining non-metals, however, involve reactions yielding products with fluorescence properties differing from those of the reactants. Thus a method for determination of NO_2 has been developed which based on the difference in the conditions for excitation and recording of the fluorescence of 5-aminofluoresceine and of its reaction product with the NO_2^- ($C_{min} = 5.10^{-5}$ μg/ml)[674, 675].

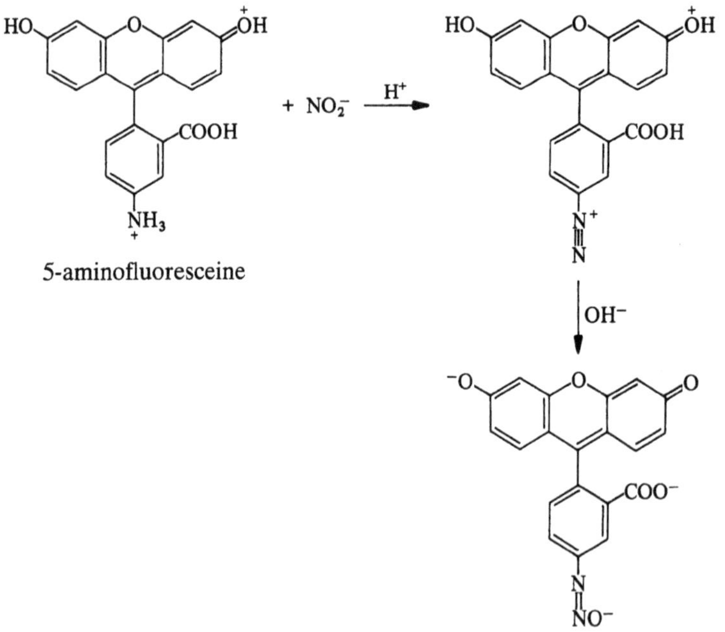

NO_2^- diazotate p-choroaniline with subsequent combination of the product with 2,6-diaminopyridine and action of ammine complexes of Cu(II). This results in the formation of a triazole ($C_{min} = 2 \times 10^{-3}$ µg/ml) which exhibits intense fluorescence in acid media[676]. For the determination of NO_2^-, a reaction has been proposed which involves oxidation of 2,3-diaminonaphthalene to fluorescent 2,3-naphthotriazole which is extracted by 1,1,2,2-tetrachloroethane ($C_{min} = 6 \times 10^{-3}$ µg/ml)[677]

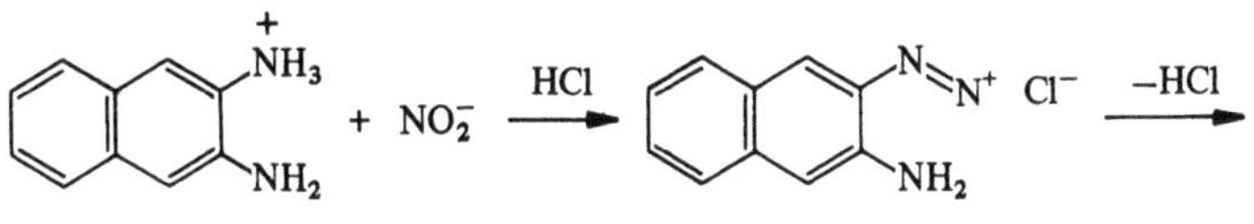

non-extractable by $C_2H_2Cl_4$

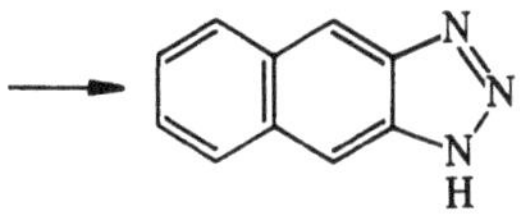

extractable by $C_2H_2Cl_4$

This reaction has been applied to the determination of the NO_3^- ($C_{min} = 1 \times 10^{-2}$ µg/ml) subsequent to its reduction by hydrazine sulfate[678]. Similar methods have been described for the determination of NO_3^- by the reaction with 2,2′-dihydroxy-4,4′-dimethoxybenzophenone[679] or 2-phenyl-1,3-benzotiazole[680], of CN^- by the reaction with benzoquinone and its 2,6-dichloro derivative[681], of HSO_3^- by the reaction with N-(p-dimethylaminophenyl)-1,4-naphthoquinonimine[682, 683].

Some methods are based on nonfluorescent products. To determine NO_3^-, its interaction with fluorsceine in H_2SO_4, which results in a nonfluorescent 2,7-dinitro derivative of the dye ($C_{min} = 1 \times 10^{-2}$ µg/ml)[684], is utilized. SO_3^{2-} are detected due to the formation of a nonfluorescent product upon reaction with formaldehyde and 5-aminofluoresceine ($C_{min} = 2 \times 10^{-2}$ µg/ml)[685].

The most selective and sensitive methods of detecting non-metals employ reactions which result in the formation of a new heterocyclic system. Thus, in the fluorimetric determination of SeO_3^{2-} this system is transformed into the strongly fluorescent 4,5-benzopiazselenole[686] or 3,4-diaminophenylpiazselanole[687] which are extracted by non-polar solvents ($C_{min} = 4 \times 10^{-4}$ and 2×10^{-3} µg/ml, respectively). These methods are at present even applied to the analysis of numerous objects[688–694]. A similar reaction is used for the detection of S^{2-}[10] yielding intensively fluorescent thionine

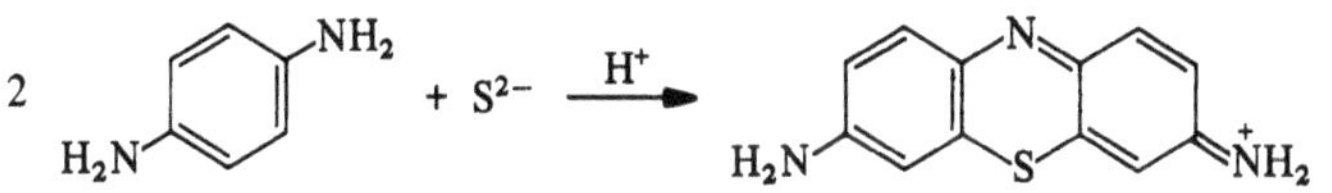

thionine

Recently, new fluorescence reactions for the determination of non-metals have been developed. A catalytic method for the detection of CN^- ($C_{min} = 3 \times 10^{-5}$ µg/ml) involving condensation reaction with p-nitrobenzaldehyde has been proposed[681]. A product of this process reacts with nonfluorescent resazurine to affort fluorescent resorufin with regeneration of the CN^-.

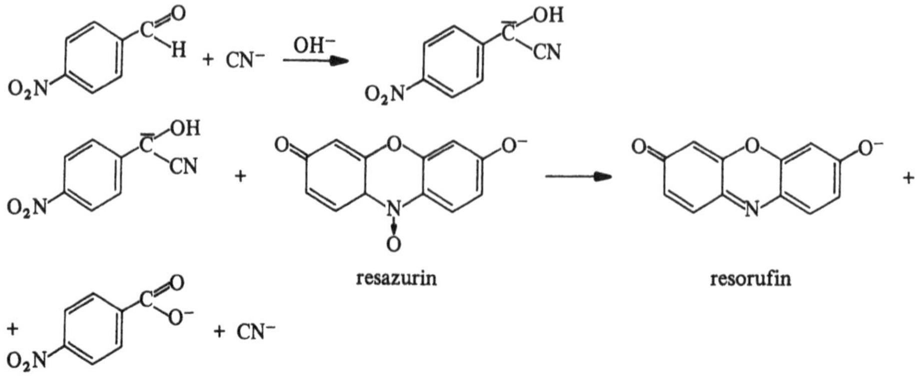

A new method involving inhibiting enzymatic transformation of homovaniline acid into fluorescent 2,2′-dihydroxy-3,3′-dimethoxydiphenyl-5,5′-diacetic acid ($C_{min} = 9 \times 10^{-1}$ μg/ml)[695] has been proposed for the determination of CN^-. This method is also useful for the determination of S^{2-} and SO_3^{2-}.

Research into the detection of anions (in particular of I^- and Br^-) utilizing fluorescence quenching of fluoresceine[696, 697], 2′,7′-bis(acetoxymercury)fluoresceine[698] and other compounds[699, 700] is being further developed. In the determination of As(V), P(V) and halide ions, extraction of the corresponding associates with rhodamine dyes[701–704], and in the detection of CN^-, S^{2-}, SO_3^{2-}, and SCN^-, chemiluminescence methods are used[705–707]. Many anions can be determined by luminescence titration which is discussed below.

E. Titrimetric Luminescence Methods

Titrimetric luminescence methods record changes in the indicator emission of radiation during titration. This change is noted visually or by instruments normally used in luminescence analysis. Most luminescence indicators are complex organic compounds which are classified into fluorescent and chemiluminescent, compounds according to the type of emission of radiation. As in titrimetry with adsorption of colored indicators, luminescence titration makes use of acid-base, precipitation, redox, and complexation reactions. Unlike color reactions, luminescence indicators enable the determination of ions in turbid or colored media and permit the detection limit to be lowered by a factor of nearly one thousand. In comparison with direct luminescence determination, the luminescence titrimetric method is more precise.

Luminescence titrimetry has been developed chiefly for acid-base titrations. Therefore, fluorescence pH-indicators are now widely used. Their application is based on changes of φ_{fl} or of the fluorescence spectrum upon the addition of a proton or its loss. At present, over 200 fluorescence pH-indicators are available; the structural formulae of the most the widely applied indicators are given in Table 8. Some of them (No. 2, 8, 9, 12, 16, 17, 23, 25 and 29) and also, primuline, tripaflavine, and rhodamine 6 G are widely used as adsorption fluorescence indicators. The titration end point can be detected in this case because of the differences in φ_{fl} of the indicator in the adsorbed state and in solution. Redox fluorescence indicators including rhodamines B and 6 G, 3,6-dihydroxyphthalic acids, complexes of Ru(II) with 2,2′-dipyridyl or 1,10-phenanthroline and other

Table 8. Fluorescence pH-indicators

No	Indicator	pH-Range	Colour of fluorescence change
1	2	3	4
1.	Benzoflavine	−0.3– 1.7	Yellow/green
2.	Eosine	0– 3.0	Non-fl./green
3.	Esculin	1.5– 2.0	Weak blue/strong blue
4.	3-Amino-1-naphthoic acid	1.5– 3.0	Non-fl./green
		4.0– 6.0	Green/blue
		11.6–13.0	Blue/non-fl.
5.	Anthranilic acid	1.5– 3.0	Non-fl./light blue
		4.5– 6.0	Light blue/dark blue
		12.5–14.0	Dark blue/non-fl.
6.	1-Naphthoic acid	2.5– 3.5	Non-fl./blue
7.	Salicylic acid	2.5– 4.0	Non-fl./dark blue
8.	Phloxin BA extra	2.5– 4.0	Non-fl./dark blue
9.	Erythrosin	2.5– 4.0	Non-fl./light green
10.	2-Naphthylamine	2.8– 4.4	Non-fl./violet
11.	3-Hydroxy-2-naphthoic acid	3.0– 6.8	Blue/green
12.	Quinine	3.0– 5.0	Blue/weak violet
		9.5–10.0	Weak violet/non-fl.
13.	Chromotropic acid	3.1– 4.4	Non-fl./light blue
14.	Morin	3.1– 4.4	Non-fl./green
		8.0– 9.8	Green/yellow-green
15.	1-Naphthylamine	3.4– 4.8	Non-fl./blue
16.	Fluorescein	4.0– 6.0	Pink-green/green
17.	Dichlorofluorescein	4.0– 6.6	Blue-green/green
18.	Resorufin	4.4– 6.4	Yellow/orange
19.	Acridine	5.2– 6.6	Green/violet
20.	5,7-Dihydroxy-4-methylcoumarin	5.5– 5.8	Light blue/dark blue
21.	2-Naphthol-6-sulfonic acid	5.7– 8.9	Non-fl./blue
22.	Quinoline	6.2– 7.2	Blue/non-fl.
23.	1-Naphthol-5-sulfonic acid	6.5– 7.5	Non-fl./green
24.	Umbelliferone	6.5– 8.0	Non-fl./blue
25.	Coumaric acid	7.2– 9.0	Non-fl./green
26.	2-Naphthol-6,8-disulfonic acid	7.5– 9.1	Blue/light blue
27.	1-Naphthol-2-sulfonic acid	8.0– 9.0	Dark blue/light blue
28.	2-Naphthol	8.5– 9.5	Non-fl./blue
29.	Acridine orange	8.4–10.4	Non-fl./yellow-green
30.	Coumarin	9.5–10.5	Non-fl./light green
31.	2-Naphthionic acid	12.0–13.0	Blue/violet
32.	2-Naphthylamine-6,8-disulfonic acid	12.0–14.0	Blue/yellow-pink

compounds whose fluorescence properties change upon oxidation or reduction are not used in analysis. The indicators in complexometric titrations are metalofluorescence indicators (Table 9) which change their fluorescence properties in complexation. Most of them are used in complexometric titration. Chemiluminescence indicators are employed for end point detection in all titrimetric methods. In contrast to fluorescence indicators, they do not need an excitation source. Chemiluminescence indicators are mainly applied in complexometric titration.

Table 9. Metallofluorescence indicators

Indicator	Metal ion	Reference
1	2	3
4-[Bis(carboxymethyl)aminomethyl]-3-hydroxy-2-naphthoic acid	Mg(II), Ca(II)	744
Bis-N,N-glycinemethylene-3,6-dichlorofluorescein	Cu(II)	714
Bis-N,N-glycinemethylene-4,5-dichlorofluorescein	Cu(II)	714
Bis-2′,7′,-N,N-glycinemethylene-4′,5′-dichlorofluorescein	Cu(II)	714
Bis-4′,5′-N,N-glycinemethylene-2′,7′-dichlorofluorescein	Cu(II)	714
Bis-N,N-glycinemethylenefluorescein	Cu(II)	714
Calcein	Ca(II)	712
Calcein blue	Mg(II), Ca(II), Fe(III)	713
Calcon	Ca(II)	739, 740
Calcon-o-anisidide	Ca(II)	739, 740
Calcon-p-anisidide	Ca(II)	739, 740
Calcon-1-naphthylamide	Ca(II)	739
Calcon-m-nitroanilide	Ca(II)	739, 740
1-[N,N-(Dicarboxymethyl)amino]-naphthaline-4-sulfonic acid	Cu(II), Fe(III)	741, 742
1-[N,N-(Dicarboxymethyl)amino]-2-naphthol-4-sulfonic acid	Cu(II), Hg(II), Co(II), Ni(II), Bi(III), Fe(III)	741, 742
1-[N,N-(Dicarboxymethyl)amino]-8-naphthol-4,6-disulfonic acid	Cu(II), Fe(III)	741, 742
Esculetin	Ca(II), Cu(II)	715
8-Hydroxyquinoline in $CHCl_3$	Al(III), Ga(III)	746
8-Hydroxyquinoline-5-sulfonic acid	Mg(II), Cu(II), Cd(II)	716
	Zn(II)	716, 717
Hydroxystilbenyl complexone	Cu(II), Hg(II), Pb(II), Mn(II), Co(II), Ni(II), Al(III), Ga(III), In(III), Sc(III), Bi(III), Fe(III)	743
Methyl calcein blue	Ca(II), Cu(II)	715
4-Methylesculetin-8-methyleneiminodiacetic acid	Ca(II), Cu(II)	715
4-Methylumbelliferrone	Ca(II), Cu(II)	715
4-Methylumbelliferrone-8-methyleneglycine	Ca(II), Cu(II)	715
Morin	Th(IV)	718, 719
	Be(II), Al(III), Ga(III), In(III), Sc(III)	719
	Zr(IV)	719, 720
Tetramercuryacetatefluorescein	S(II)	721
N,N′,N′-Tricarboxymethyl-N-1(naphthyl-4-sulfonic acid)diethylamine	Cu(II), Hg(II), Co(II) Ni(II), Bi(III), Fe(III)	741, 742
N,N′,N′-Tricarboxymethyl-N-1(2-hydroxynaphthyl-4-sulfonic acid)diethylamine	Cu(II), Hg(II), Co(II), Ni(II), Bi(III), Fe(III)	742
N,N′,N′-Tricarboxymethyl-N-1(8-hydroxynaphthyl-4,6-disulfonic acid)diethylamine	Cu(II), Hg(II), Co(II), Ni(II), Bi(III), Fe(III)	742
Umbelliferrone	Ca(II), Cu(II)	715
Umbelliferrone-8-methyleneglycine	Ca(II), Cu(II)	715
Umbelliferrone-8-methyleneiminodiacetic acid	Ca(II), Cu(II)	715
Zincone	Ca(II), Hg(II), Cd(II), Pb(II)	722

Luminescence titration has extensively been described in several monographs[4–6, 11, 708]. Fluorimetric and chemiluminescence indicators are used for the determination of about 30 elements (Fig. 3).

Research is being made in inorganic analysis of the adsorption indicators primuline[709] and tripaflavine[710, 711]; of the metalofluorescence indicators, calceine[712], calceine blue[713], and related compounds[714, 715], 8-hydroxyquinoline-5-sulfonic acid[716, 717], morin[718–720], tetramercuryacetatefluoresceine[721] and zinkone[722]; of the chemiluminescence indicators, luminole and lucigenine[723–726] and siloxene[727].

Oxazine derivatives[728], stilbene[729, 730], nitrogen bases of petroleum[731], acridone[732, 733], indole- and 5-hydroxyindole-2-carboxylic acids[734] have been studied as fluorescence pH-indicators.

The use of adsorption indicators such as 1,10-phenanthroline and eosine[735] and nitrogen bases of petroleum[736, 737] has been described.

Derivatives of coumarine[738], calcon[739, 740], complexons on the basis of naphthalene derivatives[741, 742], hydroxystilbenyl complexon[743], 4-[bis(carboxymethyl)aminomethyl]-3-hydroxy-2-naphthoic acid[744] have been suggested.

5-Hydroxyindole-2-carboxylic acid has been used as fluorimetric redox indicator[745].

A new method of extraction-fluorescence titration has been developed[746].

IV. References

1. Shcherbov, D. P., Plotnikova, R. N.: Zavodsk. Lab. *42,* 1429 (1976)
2. Konstantinova-Shlezinger, M. A. (ed.): Fluorimetric Analysis, Moscow, Gos. Izd. Fiz. – mat. Lit., 1961
3. Shcherbov, D. P.: Fluorimetry in Chemical Analysis of Mineral Raw Material, Moscow, Nedra, 1965
4. Bozhevolnov, E. A.: Luminescence Analysis of Inorganic Substances, Moscow, Khimia, 1966
5. Stolyarov, K. P., Grigorjev, N. N.: Introduction to Luminescence Analysis of Inorganic Substances, Leningrad, Khimia, 1967
6. Golovina, A. P., Levshin, L. V.: Chemical Luminescence Analysis of Inorganic Substances, Moscow, Khimia, 1978
7. Blyum, I. A.: Extraction-Photometric Methods of Analysis with Application of Basic Dyes, Moscow, Nauka, 1970
8. Poluektov, N. S., Efryushina, N. P., Gava, S. A.: Determination of Microamounts of Lanthanides by Luminescence of Crystalophosphors, Kiev, Naukova Dumka, 1976
9. Hercules, D. M. (ed.): Fluorescence and Phosphorescence Analysis. Principles and Applications, New York, Interscience, 1966
10. Parker, C. A.: Photoluminescence of Solutions with Applications to Photochemistry and Analytical Chemistry, New York, Elsevier, 1968
11. White, C. E., Argauer, R. J.: Fluorescence Analysis. A Practical Approach, New York, Dekker, 1970
12. Winefordner, J. D., Schulman, S. G., O'Haver, T. C.: Luminescence Spectrometry in Analytical Chemistry, New York, Wiley-Interscience, 1972
13. Winefordner, J. D. (ed.): Trace Analysis. Spectroscopic Methods for Elements, New York, Wiley-Interscience, 1976
14. Shcherbov, D. P.: Zavodsk. Lab. *34,* 641 (1968)
15. Shcherbov, D. P., Plotnikova, R. N.: ibid. *41,* 129 (1975)
16. White, C. E., Weissler, A.: Anal. Chem. *42,* 57 R (1970)
17. White, C. E., Weissler, A.: ibid. *44,* 182 R (1972)
18. Weissler, A.: ibid. *46,* 500 R (1974)
19. O'Donnel, C. M., Solie, T. N.: ibid. *48,* 175 R (1976)

20. O'Donnel, C. M., Solie, T. N.: ibid. *50,* 189 R (1978)
21. Bozhevolnov, E. A., Solovjev, E. A., in: Probl. Anal. Khim., Sovr. Metodi Analiza Mikroobjektov i Tonkikh Plenok, N 4, Moskva, Nauka, 1977, p. 100
22. Grigorjev, N. N., Ioannu, P. Kh., Stolyarov, K. P., in: Probl. Sovr. Anal. Khim., Spectr. Metodi Opred. Zagryazn. v Okr. Srede, N 2, Leningrad, Izd. Leningrad. Univ., 1977, p. 110
23. Favorskaya, L. V., Shcherbov, D. P. (eds.): Issled. v Obl. Khim. i Fiz. Metodov Analiza Mineral. Syrya, N 6, Alma-Ata, 1977
24. Sobalik, Z., Holzbecher, Z.: Chem. Listy *72,* 706 (1978)
25. Shcherbov, D. P.: Zavodsk. Lab. *24,* 1203 (1958)
26. Bowen, E. J. (ed.): Luminescence in Chemistry, Princeton, Van Nostrand, 1968
27. McGlynn, S. P., Azumi, T., Kinoshita, S.: Molecular Spectroscopy of the Triplet State, Englewood Cliffs, Prentice-Hall, 1969
28. Lim, E. C. (ed.): Molecular Luminescence, New York, Benjamin, 1969
29. Becker, R. S.: Theory and Interpretation of Fluorescence and Phosphorescence, New York, Wiley-Interscience, 1969
30. Udenfriend, S.: Fluorescence Assay in Biology and Medicine, New York, Academic, 1970
31. Pesce, A. J., Rosen, C. G., Pasby, T. L.: Fluorescence Spectroscopy, Introduction for Biology and Medicine, New York, Dekker, 1971
32. Winefordner, J. D. (ed.): Spectrochemical Methods of Analysis, Quantitative Analysis of Atoms and Molecules, Adv. Analyt. Chem. Instrument., Vol. 9, New York, Wiley-Interscience, 1971
33. Mavrodineanu, R., Shultz, G. I., Menis, O. (eds.): Accuracy in Spectrophotometry and Luminescence Measurements. Nat. Bur. Stand., Spec. Publ., Vol. 378, Washington, D. C., 1973
34. Guilbault, G. G.: Practical Fluorescence, Theory, Methods and Techniques, New York, Dekker, 1973
35. Pesez, M., Bartos, J.: Colorimetric and Fluorimetric Determination of Organic Compounds, New York, Dekker, 1974
36. Barltrop, J. A., Coyle, J. D.: Excited States in Organic Chemistry, New York, Wiley, 1975
37. Barnes, R. M. (ed.): Emission Spectroscopy, Stroudsberg, Dowden, Hutchinson and Ross, 1976
38. Birks, J. B.: Excited States of Biological Molecules, Chichester, Wiley, 1976
39. Mielenz, K. D., Velapoldi, R. A., Mavrodineanu, R. (eds.): Standardization in Spectrophotometry and Luminescence Measurements. Nat. Bur. Stand., Spec. Publ., Vol. 466, Washington, D. C., 1977
40. Schulman, S. G.: Fluorescence and Phosphorescence Spectroscopy, Physicochemical Principles and Practice, Oxford, Pergamon, 1977
41. McGlynn, S., Azumi, T., Kinoshita, S.: Molecular Spectroscopy of the Triplet State, Moscow, Mir, 1972
42. Barltrop, J., Coyle, J.: Excited States in Organic Chemistry, Moscow, Mir, 1978
43. Krasnovsky, A. A., Neporent, B. S. (eds.): Molecular Photonics, Leningrad, Nauka, 1970
44. Nurmukhametov, R. N.: Absorption and Luminescence of Aromatic Compounds, Moscow, Khimia, 1971
45. Terenin, A. N.: Selected Works. Elementary Processes in Complex Organic Molecules, Vol. 2, Leningrad, Nauka, 1974
46. Krasovitsky, B. M., Bolotin, B. M.: Organic Luminophores, Leningrad, Khimia, 1976.
47. Bagdasarjan, Kh. S.: Two Quantum Photochemistry, Moscow, Nauka, 1976
48. Terenin, A. N.: Photonics of Dye Molecules and Relative Organic Compounds, Leningrad, Nauka, 1967
49. Parker, C.: Photoluminescence of Solutions, Moscow, Mir, 1972
50. Winefordner, J. (ed.): Spectroscopic Methods of Determining Trace Elements, Moscow, Mir, 1979
51. Macdonald, G. L.: Anal. Chem. *50,* 135 R (1978)
52. Hieftje, G. M., Copeland, T. R.: ibid. *50,* 300 R (1978)
53. Kazankin, O. N. et al.: Inorganic Luminophores, Leningrad, Khimia, 1975
54. Fok, M. V.: Introduction to Kinetics of Crystalophosphors Luminescence, Moscow, Nauka, 1964

55. Antonov-Romanovsky, V. V.: Kinetics of Crystalophosphors Photoluminescence, Moscow, Nauka, 1966
56. Gurvitch, A. M.: Introduction to Physical Chemistry of Crystalophosphors, Moscow, Vysshaya Shkola, 1971
57. Vavilov, S. I.: Collected Works. Works in Physics in 1914–1936, Vol. 1, Moscow, Izd. AN SSSR, 1954
58. Vavilov, S. I.: Collected Works. Works in Physics in 1937–1951, Vol. 2, Moscow, Izd. AN SSSR, 1952
59. Parker, C. A., Rees, W. T.: Analyst *85,* 587 (1960)
60. Calvert, J. G., Pitts, J. N.: Photochemistry, New York, Wiley, 1966
61. Crosby, G. A., Demas, J. N., Callis, J. B.: J. Res. Nat. Bur. Stand., *76A,* 561 (1972)
62. Bennett, R. G., McCartin, P. J.: J. Chem. Phys. *44,* 1969 (1966)
63. Stevens, B., Thomaz, M. F.: Chem. Phys. Lett. *1,* 549 (1968)
64. Stacy, W. T., Swenberg, C. E.: J. Chem. Phys. *52,* 1962 (1970)
65. Hammes, G. G. (ed.): Investigation of Rates and Mechanisms of Reactions. Techniques of Chemistry, Vol. 6, Pt. 2. Investigation of Elementary Reaction Steps in Solutions and Very Fast Reactions, Chichester, Wiley, 1974
66. Emanuel, N. M., Knorre, D. G.: Chemical Kinetics, Moscow, Vysshaya Shkola, 1974
67. Bowen, E. J., Wokes, F.: Fluorescence of Solutions, London, Longmans and Green, 1953, p. 25
68. Forster, T.: Fluoreszenz organischer Verbindungen, Göttingen, Vandenhoeck and Ruprecht, 1951, p. 158
69. Birks, J. B., Dyson, D. J.: Proc. Roy. Soc. *275A,* 135 (1963)
70. Lamola, A. A., Hammond, G. S.: J. Chem. Phys. *43,* 2129 (1965)
71. Bowers, P. G., Porter, G.: Proc. Roy. Soc. *296A,* 435 (1967)
72. Bowers, P. G., Porter, G.: ibid. *299A,* 348 (1967)
73. Pavlopoulos, T. G.: J. Opt. Soc. Amer. *63,* 180 (1973)
74. Medinger, T., Wilkinson, F.: Trans. Faraday Soc. *61,* 620 (1965)
75. Sandros, K.: Acta Chem. Scand. *23,* 2815 (1969)
76. Bonnier, J. M., Jardon, P.: J. Chim. Phys. Physicochim. Biol. *67,* 577 (1970)
77. Parker, C. A., Joyce, T. A.: Trans. Faraday Soc. *62,* 2785 (1966)
78. Levshin, V. L.: Photoluminescence of Liquid and Solid Substances, Moscow, Gos. Izd. Tekhn.-teoret. Lit., 1951
79. Stepanov, B. I.: Luminescence of Complex Molecules, Vol. 2, Minsk, Izd. AN BSSR, 1955
80. Stokes, G. G.: Phil. Trans. *143,* 463 (1852)
81. Lommel, E.: Ann. Phys. *8,* 246 (1879)
82. Fennel, R. W., West, T. S.: Pure Appl. Chem. *18,* 439 (1969)
83. IUPAC-Commission, V-4, Reprint, Part II, 1974
84. Zh. Anal. Khim.*26,* 1021 (1971)
85. Ibid. *30,* 2058 (1975)
86. Shigorin, D. N.: Zh. Fiz. Khim. *44,* 2681 (1970)
87. Shigorin, D. N.: Zh. Vses. Khim. Obshch. im D. I. Mendeleeva *20,* 32 (1975)
88. Shigorin, D. N.: Zh. Fiz. Khim. *54,* 1905, 1920 (1980)
89. Shigorin, D. N.: ibid. *51,* 1894 (1977)
90. Komarov, V. M., Plotnikov, V. G.: Teor. Exper. Khim. *10,* 62 (1974)
91. Kasha, M.: Radiat. Res., Suppl. *2,* 243 (1960)
92. Nurmukhametov, R. N., Plotnikov, V. G., Shigorin, D. N.: Zh. Fiz. Khim. *40,* 1154 (1966)
93. Nurmukhametov, R. N.: Uspekhi Khim. *36,* 1629 (1967)
94. Louer, S., El-Sayed, M.: Uspekhi Fiz. Nauk *94,* 289 (1968)
95. Shigorin, D. N., Plotnikov, V. G.: Dokl. AN SSSR *234,* 121 (1977)
96. Shpolsky, E. V., Girdzhiyauskaite, E. A., Klimova, L. A. in: Mat. 10 Vses. Soveshch. po Spectr., Vol. 1, L'vov, Izd. L'vov. Univ., 1957, p. 24
97. Shigorin, D. N. et al.: Dokl. AN SSSR *120,* 1242 (1958)
98. Nurmukhametov, R. N., Shigorin, D. N., Dokunikhin, N. S.: Zh. Fiz. Khim. *34,* 2055 (1960)
99. Nurmukhametov, R. N., Shigorin, D. N., Dokunikhin, N. S.: Izv. AN SSSR. Ser. Fiz. *24,* 728 (1960)
100. Nurmukhametov, R. N., Shigorin, D. N.: Zh. Fiz. Khim. *35,* 72 (1961)

101. Shcheglova, N. A., Shigorin, D. N., Dokunikhin, N. S.: ibid. *40,* 1048 (1966)
102. Strokach, N. S., Gastilovich, E. A., Shigorin, D. N.: Dokl. AN SSSR *202,* 136 (1972)
103. Strokach, N. S., Gastilovich, E. A., Shigorin, D. N.: Opt. i Spectr. *35,* 238 (1973)
104. Gastilovich, E. A., Kryuchkova, G. T., Shigorin, D. N.: ibid. *38,* 500 (1975)
105. Kryuchkova, G. T., Gastilovich, E. A., Shigorin, D. N.: Zh. Fiz. Khim. *50,* 2678 (1976)
106. Tskhai, K. V., Gastilovich, E. A., Shigorin, D. N.: ibid. *50,* 2694 (1976)
107. Gastilovich, E. A., Tskhai, K. V., Shigorin, D. N.: Opt. i Spectr. *41,* 566 (1976)
108. Shpolsky, E. V.: Uspekhi Fiz. Nauk. *71,* 215 (1960)
109. Shpolsky, E. V.: ibid. *77,* 321 (1962)
110. Shpolsky, E. V.: ibid. *80,* 255 (1963)
111. Shcheglova, N. A., Shigorin, D. N.: Zh. Fiz. Khim. *38,* 1261 (1964)
112. Shcheglova, N. A., Shigorin, D. N., Gorelik, M. V.: ibid. *39,* 893 (1965)
113. Shigorin, D. N. et al.: ibid. *45,* 511 (1971)
114. Shcheglova, N. A. et al.: ibid. *45,* 518 (1971)
115. Ermolaev, V. L.: Opt. i Spectr. *11,* 492 (1961)
116. Nurmukhametov, R. N., Gobov, G. V.: ibid. *18,* 227 (1965)
117. Nurmukhametov, R. N. et al.: Zh. Fiz. Khim. *40,* 2206 (1966)
118. Nurmukhametov, R. N., Belaits, I. L., Shigorin, D. N.: ibid. *41,* 1928 (1967)
119. Belaits, I. L., Nurmukhametov, R. N., Shigorin, D. N.: ibid. *43,* 869 (1969)
120. Bhatnagar, D. C., Forster, L. S.: Spectrochim. Acta *21,* 1803 (1965)
121. Craven, T. L., Lytle, F. E.: Anal. Chim. Acta *107,* 273 (1979)
122. Bozhevolnov, E. A., Solovjev, E. A.: Dokl. AN SSSR *148,* 335 (1963)
123. Desai, S. R., Sudhalatha, K. K.: Talanta *14,* 1346 (1967)
124. Ryan, D. E., Snape, F., Winpe, M.: Anal. Chim. Acta *58,* 101 (1972)
125. Sommer, L., Maung-Gyee, W. P., Ryan, D. E.: Ser. Fac. Sci. Natur., Univ. Purkynianae Brun. *2,* 115 (1972)
126. Lever, M.: Anal. Chim. Acta *65,*311 (1973)
127. Taniguchi, H., Tsuge, K., Nakano, S.: Yakugaku Zasshi *94,* 759 (1974)
128. Nishikawa, Y. et al.: Bunseki Kagaku *26,* 365 (1977)
129. Shcherbov, D. P. et al. in: Issled. v Obl. Khim. i Fiz. Metodov Analiza Mineral. Syrya, N 6, Alma-Ata, 1977, p. 16
130. Chelnokova, M. N. in: Khim. i Khim. Tekhnol., Perm, Perm. Univ., 1978, p. 148
131. Karyakin, A. V. et al.: Dokl. AN SSSR *241,* 617 (1978)
132. Shcheglova, N. A., Shigorin, D. N., Dokunikhin, N. S.: Zh. Fiz. Khim. *38,* 1963 (1964)
133. Shcheglova, N. A. et al.: ibid. *48,* 271 (1974)
134. Shcheglova, N. A., Shigorin, D. N., Dokunikhin, N. S.: ibid. *50,* 2320 (1976)
135. Nurmukhametov, R. N. et al. in: Fiz. Probl. Spectr., Mater. 13 Soveshch., Vol. 1, Moskva, Izd. AN SSSR, 1962, p. 289
136. Nurmukhametov, R. N. et al.: Opt. i Spectr. *11,* 606 (1961)
137. Nurmukhametov, R. N. et al.: Dokl. AN SSSR *143,* 1145 (1962)
138. Kozlov, Yu. I. et al.: Zh. Fiz. Khim. *37,* 2432 (1963)
139. Krasovitsky, B. M., Smelyakova, V. B., Nurmukhametov, R. N.: Opt. i Spectr. *17,* 558 (1964)
140. Alimarin, I. P., Golovina, A. P., Runov, V. K.: Izv. AN SSSR, Ser. Khim. N 6, 1423 (1974)
141. Alimarin, I. P., Golovina, A. P., Runov, V. K.: Rev. Roum. Chim. *22,* 503 (1977)
142. Bryukhanov, V. V., Runov, V. K., Golovina, A. P. in: Tez. Dokl. 26 Vses. Soveshch. po Lumin., Samarkand, 1979, p. 70
143. Lisitsina, D. N., Shcherbov, D. P.: Zh. Anal. Khim. *25,* 2310 (1970)
144. Lisitsina, D. N., Shcherbov, D. P.: ibid. *28,* 1203 (1973)
145. Shcherbov, D. P., Lisitsina, D. N. in: Issled. v Obl. Khim. i Fiz. Metodov Analiza Mineral. Syrya, N 4, Alma-Ata, 1975, p. 62
146. Mori, I., Enoki, T., Mano, T.: Bunseki Kagaku *22,* 1202 (1973)
147. Alykov, N. M., Grunin, A. V.: Isv. Vyssh. Uchebn. Zaved. Ser. Khim. i Technol. *17,* 1254 (1974)
148. Maksimycheva, Z. T., Talipov, Sh. T., Artemova, V. Ya.: Nauch. Tr. Tashkent. Univ., Voprosi Khim., N 435, Tashkent, 1973, p. 39
149. Morisige, K. et al.: Bunseki Kagaku *27,* 109 (1978)
150. Deguchi, M. et al.: ibid. *28,* 127 (1979)

151. Drevenkar, V., Stefanac, Z., Brbot, A.: Microchem. J. *21,* 402 (1976)
152. Dolgorev, A. V., Serikov, Yu. A., Zolotavin, V. L.: Zh. Anal. Khim. *33,* 2357 (1978)
153. Dolgorev, A. V., Serikov, Yu. A., Chesnokova, N. A.: Zavodsk. Lab. *45,* 691 (1979)
154. Uno, T., Taniguchi, H.: Bunseki Kagaku *20,* 1123 (1971)
155. Hydes, D. J., Liss, P. S.: Analyst *101,* 922 (1976)
156. Hiraki, K.: Bull. Chem. Soc. Jap. *45,* 789 (1972)
157. Shakirova, T. A. et al. in: Nauch. Tr. Tashkent. Univ., Voprosi Khim., N 435, Tashkent, 1973, p. 68
158. Dzhiyanbaeva, R. Kh. et al. in: Nauch. Tr. Tashkent. Univ., Voprosi Khim., N 419, Tashkent, 1972, p. 84
159. Pilipenko, A. T., Zhebentyaev, A. I.: Zh. Anal. Khim. *33,* 2119 (1978)
160. Chernova, R. K. et al.: ibid. *31,* 37 (1976)
161. Hiraki, K.: Bull. Chem. Soc. Jap. *45,* 1395 (1972)
162. Blyum, I. A., Pavlova, N. N., Kalupina, F. P.: Zh. Anal. Khim. *26,* 55 (1971)
163. Podberezskaya, N. K., Shilenko, E. A., Shcherbov, D. P.: Zavodsk. Lab. *36,* 661 (1970)
164. Lapid, J., Farhi, S., Koresh, Y.: Anal. Lett. *9,* 355 (1976)
165. Pszonicki, L., Tkacz, W.: Anal. Chim. Acta *87,* 177 (1976)
166. Tkacz, W., Pszonicki, L.: Chem. Anal. (PRL) *22,* 801 (1977)
167. Tkacz, W., Pszonicki, L.: ibid. *22,* 1013 (1977)
168. Tkacz, W., Pszonicki, L.: Anal. Chim. Acta *90,* 339 (1977)
169. Alykov, N. M., Cherkesov, A. I.: Zh. Anal. Khim. *31,* 1104 (1976)
170. Savvin, S. B., Chernova, R. K., Kudryavtseva, L. M.: ibid. *31,* 269 (1976)
171. Ryan, D. E., Granda, M., Janmohammed, M.: Anal. Chim. Acta *76,* 467 (1975)
172. Murata, A., Tominaga, M., Suzuki, T.: Bunseki Kagaku *23,* 1349 (1974)
173. Ito, T., Murata, A.: ibid. *20,* 335 (1971)
174. Capitan, F., Salinas, F., Franquelo, L. M.: Anal. Lett. *8,* 753 (1975)
175. Talipov, Sh. T. et al.: Dokl. AN UzSSR, N 6, 36 (1974)
176. Gladilovich, D. B., Grigorjev, N. N., Stolyarov, K. P.: Zh. Anal. Khim. *33,* 2113 (1978)
177. Bishop, J. A.: Anal. Chim. Acta *87,* 255 (1976)
178. Morisige, K.: ibid. *73,* 245 (1974)
179. Murata, A., Nakamura, M.: Bunseki Kagaku *21,* 1365 (1972)
180. Kabrt, L., Holzbecher, Z.: Coll. Czech. Chem. Comm. *41,* 540 (1976)
181. Guiraum, A., Vilchez, J. L.: Quim. Anal. (pura y apl.) *29,* 265 (1975)
182. Tashkhodzhaev, A. T. et al.: Zavodsk. Lab. *41,* 280 (1975)
183. Tashkhodzhaev, A. T. et al.: Uzb. Khim. Zh. N 6, 9 (1975)
184. Talipov, Sh. T. et al. in: Nauch. Tr. Tashkent. Univ., Voprosi Khim., N 419, Tashkent, 1972, p. 89
185. Talipov, Sh. T. et al.: Zh. Anal. Khim. *28,* 807 (1973)
186. Holzbecher, Z., Volka, K.: Coll. Czech. Chem. Comm. *35,* 2925 (1970)
187. Alykova, T. V., Cherkesov, A. I., Alykov, N. M.: Isv. Vyssh. Uchebn. Zaved. Ser. Khim. i Khim. Tekhnol. *15,* 1107 (1972)
188. Beltyukova, S. V. et al.: Ukr. Khim. Zh. *42,* 83 (1976)
189. Hefley, A. J., Jaselskis, B.: Anal. Chem. *46,* 2036 (1974)
190. Stolyarov, K. P., Firyulina, V. V.: Zh. Anal. Khim. *33,* 2102 (1978)
191. Matveets, M. A., Shcherbov, D. P.: ibid. *26,* 823 (1971)
192. Kasa, I., Bajnoczy, G.: Period. Polytechn. Chem. Eng. *18,* 289 (1974)
193. Bozhevolnov, E. A. et al.: Zh. Anal. Khim. *25,* 1722 (1970)
194. Watanabe, K., Kawagaki, K.: Bunseki Kagaku *23,* 1356 (1974)
195. Nagasawa, K., Ishidaka, O.: Chem. Pharm. Bull. *22,* 375 (1974)
196. Pal, B. K. et al.: Anal. Chim. Acta *88,* 353 (1977)
197. Chan, Ti Khjeu, Getman, T. E., Volkova, A. I.: Ukr. Khim. Zh. *36,* 87 (1970)
198. Haddad, P. R., Alexander, P. W., Smythe, L. E.: Talanta *23,* 275 (1976)
199. Pilipenko, A. T., Shevchenko, T. L., Volkova, A. I.: Zh. Anal. Khim. *32,* 731 (1977)
200. Koh, K. J., Ryan, D. E.: Anal. Chim. Acta *52,* 503 (1970)
201. Koh, K. J., Ryan, D. E.: ibid. *54,* 303 (1971)
202. Grigoryan, L. A., Pogosyan, A. N., Tarayan, V. M.: Arm. Khim. Zh. *25,* 931 (1972)

203. Akhmedli, M. K., Efendiev, D. A., Ruvinova, F. I.: Uch. Zap. Azerb. Univ. Ser. Khim. Nauk, N4, 10 (1973)
204. Maksimycheva, Z. T. et al.: Izv. Vyssh. Uchebn. Zaved. Ser. Khim. i Khim. Tekhnol. *17,* 348 (1974)
205. Tashkhodzhaev, A. T. et al.: Zh. Vses. Khim. Obshch. im. D. I. Mendeleeva *21,* 114 (1976)
206. Kina, K., Hirokata, K., Ishibashi, N.: Bunseki Kagaku *26,* 246 (1977)
207. Kina, K., Shiraishi, K., Ishibashi, N.: ibid. *25,* 501 (1976)
208. Pilipenko, A. T. et al.: Ukr. Khim. Zh. *43,* 536 (1977)
209. Kato, H. et al.: Bunseki Kagaku *20,* 1315 (1971)
210. Olenovich, N. L., Kovalchuk, L. I.: Zh. Anal. Khim. *28,* 2162 (1973)
211. Talipov, Sh. T. et al.: Izv. Vyssh. Uchebn. Zaved. Ser. Khim. i Khim. Tekhnol. *16,* 1154 (1973)
212. Watanabe, K., Kawagaki, K.: Bull. Chem. Soc. Jap. *48,* 1812 (1975)
213. Tashkhodzhaev, A. T. et al.: Zavodsk. Lab. *41,* 281 (1975)
214. Golovina, A. P. et al.: Zh. Anal. Khim. *25,* 2242 (1970)
215. Talipov, Sh. T. et al.: Izv. Vyssh. Uchebn. Zaved. Ser. Khim. i Khim. Tekhnol. *15,* 1109 (1972)
216. Pilipenko, A. T. et al.: Ukr. Khim. Zh. *37,* 689 (1971)
217. Flyantikova, G. V., Korolenko, L. I.: Zh. Anal. Khim. *30,* 1349 (1975)
218. Capitan, F., Roman, M., Fernandez-Gutierrez, A.: Bol. Soc. Quim. Peru *40,* 65 (1974)
219. Talipov, Sh. T. et al.: Zavodsk. Lab. *44,* 1052 (1978)
220. Pilipenko, A. T. et al.: Zh. Anal. Khim. *27,* 1787 (1972)
221. Pilipenko, A. T., Kukibaev, T. U., Volkova, A. I.: ibid. *28,* 510 (1973)
222. Kulikov, N. S., Granovsky, Yu. V., Golovina, A. P.: Izv. Vyssh. Uchebn. Zaved. Ser. Khim. i Khim. Tekhnol. *16,* 1006 (1973)
223. Pilipenko, A. T., Kukibaev, T. U., Volkova, A. I.: Zh. Anal. Khim. *29,* 710 (1974)
224. Blyum, I. A., Brushtein, N. A., Oparina, L. I.: ibid. *26,* 48 (1971)
225. Oshima, G., Nagasawa, K.: Chem. Pharm. Bull. *18,* 687 (1970)
226. Holzbecher, J., Ryan, D. E.: Anal. Chim. Acta *64,* 333 (1973)
227. Yamane, Y. et al.: Bunseki Kagaku *22,* 192 (1973)
228. Titkov, Yu. B.: Ukr. Khim. Zh. *37,* 502 (1971)
229. Pilipenko, A. T., Zhebentyaev, A. I., Volkova, A. I.: ibid. *42,* 998 (1976)
230. Watanabe, K., Fujiwara, A., Kawagaki, K.: Bull. Chem. Soc. Jap. *50,* 1460 (1977)
231. Pilipenko, A. T., Zhebentyaev, A. I.: Ukr. Khim. Zh. *43,* 1314 (1977)
232. Kina, K., Shiraishi, K., Ishibashi, N.: Bunseki Kagaku *27,* 291 (1978)
233. Poluektov, N. S., Meshkova, S. B., Melentjeva, E. V.: Zh. Anal. Khim. *25,* 1314 (1970)
234. Poluektov, N. S., Lauer, R. S., Gaidarzhi, O. F.: ibid. *26,* 898 (1971)
235. Hayashi, T., Kawai, S., Ohno, T.: Chem. Pharm. Bull. *21,* 1147 (1973)
236. Korkuc, A., Lesz, K.: Chem. Anal. (PRL) *17,* 855 (1972)
237. Pilipenko, A. T. et al.: Ukr. Khim. Zh. *43,* 752 (1977)
238. Titkov, Yu. B.: ibid. *36,* 613 (1970)
239. Pilipenko, A. T., Zhebentyaev, A. I., Volkova, A. I.: ibid. *41,* 262 (1975)
240. Haddad, P. R., Alexander, P. W., Smythe, L. E.: Talanta *22,* 61 (1975)
241. Pilipenko, A. T., Volkova, A. I., Zhebentyaev, A. I.: Zh. Anal. Khim. *26,* 2048 (1971)
242. Volkova, A. I., Getman, T. E.: Ukr. Khim. Zh. *37,* 53 (1971)
243. Pilipenko, A. T., Zhebentyaev, A. I., Volkova, A. I.: ibid. *38,* 363 (1972)
244. Pilipenko, A. T., Volkova, A. I., Zhebentyaev, A. I.: Zh. Anal. Khim. *26,* 117 (1971)
245. Shiryaev, P. A., Rigin, V. I.: Zavodsk. Lab. *41,* 917 (1975)
246. Shcherbov, D. P. et al.: Zh. Anal. Khim. *32,* 1932 (1977)
247. Grigoryan, L. A., Gaibakyan, A. G., Tarayan, V. M.: Zavodsk. Lab. *40,* 136 (1974)
248. Pilipenko, A. T., Volkova, A. I., Shevchenko, T. L.: Ukr. Khim. Zh. *41,* 1190 (1975)
249. Grigoryan, L. A., Lebedeva, S. P., Tarayan, V. M.: Arm. Khim. Zh. *28,* 540 (1975)
250. Pilipenko, A. T., Volkova, A. I., Shevchenko, T. L.: Zh. Anal. Khim. *28,* 1524 (1973)
251. Blyum, I. A., Kalupina, F. P., Tsenskaya, T. I.: ibid. *29,* 1572 (1974)
252. Filer, T. D.: Anal. Chem. *43,* 725 (1971)
253. Dolgorev, A. V., Pavlova, N. N., Ershova, V. A.: Zavodsk. Lab. *39,* 658 (1973)
254. Dolgorev, A. V. et al.: Zh. Anal. Khim. *34,* 106 (1979)

255. Morisige, K. et al.: Bunseki Kagaku *24,* 321 (1975)
256. Matveets, M. A., Akhmetova, S. D., Shcherbov, D. P.: Zh. Anal. Khim. *32,* 2143 (1977)
257. Murata, A. et al.: Bunseki Kagaku *27,* 788 (1978)
258. Nakamura, M., Murata, A.: ibid. *22,* 1474 (1973)
259. Temkina, V. Ya. et al.: Zh. Anal. Khim. *25,* 894 (1970)
260. Zeltser, L. E. et al.: ibid. *34,* 896 (1979)
261. Beltyukova, S. V. et al.: ibid. *30,* 1321 (1975)
262. Kononenko, L. I. et al.: ibid. *30,* 1716 (1975)
263. Beltyukova, S. V., Drobyazko, V. N., Kononenko, L. I.: Ukr. Khim. Zh. *43,* 641 (1977)
264. Shumova, T. I. in: Metodi Khim. Analiza Mineral. Syrya, N 14, Moskva, 1975, p. 51
265. Mori, I., Fujita, Y., Enoki, T.: Bunseki Kagaku *25,* 388 (1976)
266. Pilipenko, A. T., Lisichenok, S. L., Volkova, A. I.: Ukr. Khim. Zh. *42,* 976 (1976)
267. Nishikawa, Y. et al.: Bunseki Kagaku *19,* 1224 (1970)
268. Filer, T. D.: Anal. Chem. *43,* 1753 (1971)
269. Blyum, I. A., Pronkina, G. T., Shumova, T. I.: Zh. Anal. Khim. *25, 511* (1970)
270. Pilipenko, A. T., Volkova, A. I., Zhebentyaev, A. I.: Ukr. Khim. Zh. *37,* 578 (1971)
271. Bovay, M., Marcantonatos, M.: Anal. Chim. Acta *80,* 180 (1975)
272. Granovsky, Yu. V. et al.: Zh. Anal. Khim. *29,* 1959 (1974)
273. Morozova, L. A. et al.: Izv. Vyssh. Uchebn. Zaved. Ser. Khim. i Khim. Tekhnol. *21,* 672 (1978)
274. Filer, T. D.: Anal. Chem. *42,* 1265 (1970)
275. Grigoryan, L. A., Mirzoyan, F. V., Tarayan, V. M.: Zh. Anal. Khim. *28,* 1962 (1973)
276. Grigoryan, L. A., Artsruni, V. Zh., Tarayan, V. M.: Arm. Khim. Zh. *27,* 188 (1974)
277. Tarayan, V. M. et al.: ibid. *26,* 996 (1973)
278. Blyum, I. A., Chuvileva, A. I.: Zh. Anal. Khim. *25,* 18 (1970)
279. Koh, K. J., Ryan, D. E.: Anal. Chim. Acta *57,* 295 (1971)
280. Roman, M., Garcia-Sanchez, F., Gomez-Hens, A.: Quim. Anal. (pura y apl.) *28,* 191 (1974)
281. Pilipenko, A. T., Zhebentyaev, A. I., Volkova, A. I.: Zh. Anal. Khim. *29,* 1854 (1974)
282. Pilipenko, A. T., Zhebentyaev, A. I., Volkova, A. I.: Ukr. Khim. Zh. *41,* 1087 (1975)
283. Bozhevolnov, E. A. et al. in: Org. Reag. v Anal. Khim., Tez. 4 Vses. Konf., Vol. 2, Kiev, Naukova Dumka, 1976, p. 84
284. Kachin, S. V. et al. in: Reakt. i Osobo Chist. Veshch., N 5, Moskva, Nauch.-issled. Inst. Tekhn.-Ekon. Issled., 1978, p. 19
285. Kato, H. et al.: Bunseki Kagaku *21,* 856 (1972)
286. Ito, T., Murata, A.: ibid. *23,* 274 (1974)
287. Tashkhodzhaev, A. T. et al.: Dokl. AN UzSSR, N *9,* 26 (1973)
288. Schneider, H. O., Roselli, M. E.: Analyst *96,* 330 (1971)
289. Filer, T. D.: Anal. Chem. *43,* 469 (1971)
290. Kirkbright, G. F., Thompson, J. V., West, T. S.: ibid. *42,* 782 (1970)
291. Solovjev, E. A. et al.: Zh. Anal. Khim. *24,* 231 (1969)
292. Solovjev, E. A., Lebedeva, N. A. in: Novye Metodi Khim. Analiza Mater., Moskva, Mosk. Dom Nauch.-tekhn. Propagandi im F. E. Dzerzhinskogo, 1971, p. 103
293. Solovjev, E. A., Bozhevolnov, E. A. in: Metodi Analiza Khim. Reakt. i Preparatov, Lumin. Metodi Opred. Mikrokol. Elementov, N 11, Moskva, IREA, 1965, p. 21
294. Bozhevolnov, E. A., Solovjev, E. A. in: Sovremen. Metodi Analiza, Metodi Issled Khim. Sost. i Stroen. Veshch., Moskva, Nauka, 1965, p. 75
295. Bozhevolnov, E. A., Solovjev, E. A. in: Priklad. Spectr., Mater. *16* Soveshch., Moskva, Nauka, 1969, p. 166
296. Marcantonatos, M., Gamba, G., Monnier, D.: Anal. Chim. Acta *67,* 220 (1973)
297. Holzbecher, Z., Hejtmanek, M., Sobalik, Z.: Coll. Czech. Chem. Comm. *43,* 3325 (1978)
298. Golovina, A. P., Kachin, S. V., Runov, V. K.: Vestn. Mosk. Univ. Ser. Khim. *18,* 709 (1977)
299. Meshkova, S. B. et al.: Zh. Anal. Khim. *32,* 1529 (1977)
300. Meshkova, S. B. et al. in: Stroenie, Svojstva i Primenenie β-Diketonatov, Moskva, Nauka, 1978, p. 191
301. Yuster, P., Weissman, S. I.: J. Chem. Phys. *17,* 1182 (1949)
302. Weissman, S. I.: ibid. *18,* 1258 (1950)
303. Becker, R. S., Kasha, M.: J. Amer. Chem. Soc. *77,* 3669 (1955)

304. Allison, J. B., Becker, R. S.: J. Chem. Phys. *32,* 1410 (1960)
305. Becker, R. S., Allison, J. B.: ibid. *67,* 2662 (1963)
306. Becker, R. S., Allison, J. B.: ibid. *67,* 2669 (1963)
307. Allison, J. B., Becker, R. S.: ibid. *67,* 2675 (1963)
308. Kuzin, E. L. in: Tez. Dokl. 3 Vses. Kohf. po Anal. Khim., Vol. 2, Minsk, 1979, p. 141
309. Ermolaev, V. L. et al.: Nonradiative Energy Transfer of Electronic Excitation, Leningrad, Nauka, 1977
310. Sinha, A. P. B. in: Spectroscopy in Inorganic Chemistry, Vol. 2, New York, Academic, 1971,p. 255
311. Morita, M., Shionoya, S.: Bull. Chem. Soc. Jap. *43,* 2404 (1970)
312. Syczewsky, M.: Rocz. Chem. *44,* 1623 (1970)
313. Kropp, J. L., Windsor, M. W.: J. Chem. Phys. *39,* 2769 (1963)
314. Bhaumik, M. L., Telk, C. L.: J. Opt. Soc. Amer. *54,* 1211 (1964)
315. Melby, L. R. et al.: J. Amer. Chem. Soc. *86,* 5117 (1964)
316. Filipescu, N., Sager, W. F., Serafin, F. A.: J. Phys. Chem. *68,* 3324 (1964)
317. Bhaumik, M. L.: J. Chem. Phys. *40,* 3711 (1964)
318. Kropp, J. L., Windsor, M. W.: ibid. *42,* 1599 (1965)
319. Borkowski, R. P., Forest, H., Grafstein, D.: ibid. *42,* 2974 (1965)
320. Gallagher, P. K.: ibid. *43,* 1742 (1965)
321. Charles, R. G., Riedel, E. P.: J. Inorg. Nucl. Chem. *28,* 527 (1966)
322. Heller, A.: J. Amer. Chem. Soc. *88,* 2058 (1966)
323. Ross, D. L., Blanc, J., Pressley, R. J.: Appl. Phys. Lett. *8,* 101 (1966)
324. Bhaumik, M. L., Nugent, L. J.: J. Chem. Phys. *43,* 1680 (1965)
325. Filipescu, N., McAvoy, N.: J. Inorg. Nucl. Chem. *28,* 253 (1966)
326. Syczewski, M.: Rocz. Chem. *44,* 721 (1970)
327. Meshkova, S. B. et al.: Zh. Anal. Khim. *34,* 121 (1979)
328. Wybourne, B. G.: Spectroscopic Properties of Rare Earths, New York, Interscience, 1965
329. Crosby, G. A.: Molec. Cryst. *1,* 37 (1966)
330. Poluektov, N. S., Kononenko, L. I.: Spectrophotometric Methods of Determining Individuel Rare Earth Elements, Kiev, Naukova Dumka, 1968
331. Kuznetsova, V. V. et al.: Applications of Luminescence for the Control and Analysis of Materials Containing Rare Earth Elements, Minsk, Inst. Fis. AN BSSR, 1970
332. Karyakin, A. V. et al.: Spectral Analysis of Rare Earth Oxides, Moscow, Nauka, 1974
333. Dieke, G. H., Duncan, A. B. F.: Spectroscopic Properties of Uranium Compounds, New York, McGraw-Hill, 1949
334. Rabinowitch, E., Belford, R. L.: Spectroscopy and Photochemistry of Uranyl Compounds. Internat. Ser. Monogr. Nuclear Energy, Chemistry Div., Vol. 1, New York, Macmillan, 1964
335. Dobrolyubskaya, T. S.: Luminescence Methods of Determining Uranium, Moscow, Nauka, 1968
336. Solovjev, E. A., Bozhevolnov, E. A.: Zh. Anal. Khim. *27,* 1817 (1972)
337. Fleischauer, P. D., Fleischauer, P.: Chem. Rev. *70,* 199 (1970)
338. Titov, V. I. et al.: Assortiment Org. Reakt. dlya Opred. Khroma, Ser. Reakt. i Osobo Chist. Veshch., Moskva, 1974
339. Wrighton, M., Morse, D. L.: J. Amer. Chem. Soc. *96,* 998 (1974)
340. Veening, H., Brandt, W. W.: Anal. Chem. *32,* 1426 (1960)
341. Shcherbov, D. P., Gladysheva, G. P., Ivankova, A. I.: Zavodsk. Lab. *37,* 1300 (1971)
342. Demas, J. N., Crosby, G. A.: J. Amer. Chem. Soc. *93,* 2841 (1971)
343. Watts, R. J., Crosby, G. A.:ibid. *94,* 2606 (1972)
344. Harrigan, R. W., Hager, G. D., Crosby, G. A.: Chem. Phys. Lett. *21,* 187 (1973)
345. Navon, G., Sutin, N.: Inorg. Chem. *13,* 2159 (1974)
346. Van Houten, J., Watts, R. J.: J. Amer. Chem. Soc. *97,* 3843 (1975)
347. Demas, J. N., Addington, J. W.: ibid. *98,* 5800 (1976)
348. Demas, J. N. et al.: ibid. *97,* 3838 (1975)
349. Diomedi Camassei, F., Ancarani-Rossiello, L., Castelli, F.: J. Lumin. *8,* 71 (1973)
350. De Armond, M. K., Hillis, J. E.: J. Chem. Phys. *54,* 2247 (1971)
351. Thomas, T. R., Crosby, G. A.: J. Mol. Spectr. *38,* 118 (1971)
352. Hillis, J. E., De Armond, M. K.: Chem. Phys. Lett. *10,* 325 (1971)

353. Halper, W., De Armond, M. K.: J. Lumin. *5,* 225 (1972)
354. Thomas, T. R., Watts, R. J., Crosby, G. A.: J. Chem. Phys. *59,* 2123 (1973)
355. Halper, W., De Armond, M. K.: Chem. Phys. Lett. *24,* 114 (1974)
356. Petersen, J. D., Watts, R. J., Ford, P. C.: J. Amer. Chem. Soc. *98,* 3188 (1976)
357. Watts, R. J., Crosby, G. A., Sansregret, J. L.: Inorg. Chem. *11,* 1474 (1972)
358. Flynn, C. M., Demas, J. N.: J. Amer. Chem. Soc. *96,* 1959 (1974)
359. Watts, R. J.: ibid. *96,* 6186 (1974)
360. Watts, R. J., Griffith, B. G., Harrington, J. S.: ibid. *98,* 674 (1976)
361. Eljyashevich, M. A.: Spectra of Rare Earth Elements, Moscow, Gos. Izd. Techn.-teoret. Lit., 1953
362. Dieke, G. H.: Spectra and Energy Levels of Rare Earth Ions in Cristals, New York, Interscience, 1968
363. Gajduk, M. I., Zolin, V. F., Gajgerova, L. S.: Luminescence Spectra of Europium, Moscow, Nauka, 1974
364. Sviridov, D. T., Sviridova, R. K., Smirnov, Yu. F.: Optical Spectra of Transitional Metal Ions in Crystals, Moscow, Nauka, 1976
365. Zolin, V. F., Koreneva, L. G.: Rare Earth Probing in Chemistry and Biology, Moscow, Nauka, 1980
366. Solovjev, E. A. et al.: Zh. Anal. Khim. *25,* 1342 (1970)
367. Antipenko, B. M., Batyaev, I. M., Lyubimov, E. I.: Opt. i Spectr. *33,* 938 (1972)
368. Tokousbalides, P., Chrysochoos, J.: J. Phys. Chem. *76,* 3397 (1972)
369. Chrysochoos, J., Tokousbalides, P.: Spectr. Lett. *6,* 435 (1973)
370. Bhatt, B. C., Joshi, G. C., Pant, D. D.: Indian J. Pure Appl. Phys. *11,* 226 (1973)
371. Arapova, E. Ya. et al. in: Tr. Komiss. po Anal. Khim., Metodi Opred. Primesei v Christ. Metall., Vol. 12, Moskva, Izd. AN SSSR, 1960, p. 344
372. Levshin, V. L., Arapova, E. Ya., Baranova, E. G. in: ibid., p. 393
373. Trofimov, A. K.: Izv. AN SSSR. Ser. Fiz. *25,* 460 (1961)
374. Linares, R. C.: J. Opt. Soc. Amer. *56,* 1700 (1966)
375. Karyakin, A. V., Anikina, L. I., Filatkina, L. A.: Zh. Anal. Khim. *21,* 1196 (1966)
376. Melamed, Sh. G., Antonov, A. V., Kulevsky, L. V.: Zavodsk. Lab. *33,* 712 (1967)
377. Shand, W. A.: J. Mater. Sci. *3,* 344 (1968)
378. Ozawa, L., Toryu, T.: Anal. Chem. *40,* 187 (1968)
379. Larach, S.: Anal. Chim. Acta *42,* 407 (1968)
380. Jaworowski, R. J. et al.: Spectrochim. Acta *23 B,* 751 (1968)
381. Poluektov, N. S., Gava, S. A.: Zh. Prykl. Spectr. *9,* 268 (1968)
382. Bozhevolnov, E. A., Fakeeva, O. A. in: Metodi Analiza Khim. Reakt. i Preparatov, N 15, Moskva, IREA, 1968, p. 41
383. Anikina, L. I. et al.: Zh. Anal. Khim. *24,* 1014 (1969)
384. Gava, S. A., Poluektov, N. S.: Zavodsk. Lab. *35,* 20 (1969)
385. Poluektov, N. S., Vitkun, R. A., Gava, S. A.: Zh. Anal. Khim. *24,* 693 (1969)
386. Fakeeva, O. A., Bozhevolnov, E. A. in: Metodi Analiza Khim. Reakt. i Preparatov, Fiz.-Khim. Metodi Analiza Veshch. Osoboj Chist., N 16, Moskva, IREA, 1969, p. 165
387. Karyakin, A. V., Anikina, L. I., Le V'et Bin': Zh. Anal. Khim. *24,* 1156 (1969)
388. Poluektov, N. S., Gava, S. A.: Zavodsk. Lab. *35,* 1458 (1969)
389. Reisfeld, R., Greenberg, E.: Anal. Chim. Acta *47,* 155 (1969)
390. Poluektov, N. S., Gava, S. A.: Zh. Anal. Khim. *25,* 1735 (1970)
391. Antonov, A. V., Melamed, Sh. G. in: Anal. Khim. Redkikh Metall. i Poluprovod. Mater., Moskva, Mosk. Dom Nauch.-tekhn. Propagandi im F. E. Dzerzhinskogo, 1970, p. 149
392. Voropaeva, S. V., Pisarenko, V. F.: Uch. Zap. Mosk. Ped. Inst. im V. I. Lenina, N 391, Moskva, 1970, p. 117
393. Gava, S. A., Poluektov, N. S. in: Metodi Poluch. i Analiza Veshch. Osoboj Chist., Tr. Vses. Konf., Moskva, Nauka, 1970, p. 158
394. Smirdova, N. I., Efryushina, N. P., Poluektov, N. S.: Zavodsk. Lab. *36,* 1183 (1970)
395. Poluektov, N. S., Smirdova, N. I., Efryushina, N. P.: Zh. Anal. Khim. *25,* 715 (1970)
396. Poluektov, N. S., Smirdova, N. I., Efryushina, N. P.: ibid. *25,* 1902 (1970)
397. Anikina, L. I., Karyakin, A. V., Le V'et Bin': ibid. *25,* 1731 (1970)

398. D'Silva, A. P., De Kalb, E. L., Fassel, V. A.: Anal. Chem. *42,* 1846 (1970)
399. Saranathan, T. R., Fassel, V. A., De Kalb, E. L.: ibid. *42,* 325 (1970)
400. Reisfeld, R., Biron, E.: Talanta *17,* 105 (1970)
401. Reisfeld, R., Gur-Arieh, Z., Greenberg, E.: Anal. Chim. Acta *50,* 249 (1970)
402. Reisfeld, R., Greenberg, E., Kraus, S.: ibid. *51,* 133 (1970)
403. Poluektov, N. S. et al.: Zavodsk. Lab. *37,* 266 (1971)
404. Anikina, L. I., Karyakin, A. V., Le V'et Bin': Zh. Anal. Khim. *26,* 511 (1971)
405. Antonov, V. A., Melamed, Sh. G., Kulevsky, L. V. in: Nauch. Tr. Nauch.-issled. i Proekt. Inst. Redkometall. Promysh., Metallurgiya Redkikh Metall., Vol. 42, Moskva, 1972, p. 148
406. Poluektov, N. S., Smirdova, N. I., Efryushina, N. P.: Zh. Anal. Khim. *27,* 1616 (1972)
407. Poluektov, N. S., Efryushina, N. P., Smirdova, N. I.: Ukr. Khim. Zh. *38,* 365 (1972)
408. Larach, S., Shrader, R. E.: Anal. Chim. Acta *63,* 459 (1973)
409. D'Silva, A. P., Fassel, V. A.:Anal. Chem. *45,* 542 (1973)
410. De Kalb, E. L., D'Silva, A. P., Fassel, V. A.: Izv. AN SSSR. Ser. Fiz. *37,* 790 (1973)
411. Efryushina, N. P. et al.: Zavodsk. Lab. *39,* 129 (1973)
412. Efryushina, N. P. et al.: Zh. Anal. Khim. *28,* 1213 (1973)
413. Zhikhareva, E. A., Zelyukova, Yu. V., Poluektov, N. S.: Zavodsk. Lab. *40,* 1434 (1974)
414. Zhikhareva, E. A. et al.: Zh. Anal. Khim. *29,* 76 (1974)
415. Smirdova, N. I., Efryushina, N. P., Poluektov, N. S.: ibid. *29,* 1719 (1974)
416. Townshend, A.: Proc. Soc. Anal. Chem. *11,* 179 (1974)
417. Poluektov, N. S., Efryushina, N. P. in Uspekhi Anal. Khim., Moskva, Nauka, 1974, p. 64
418. Jayawant, D. V., Iyer, N. S., Murthy, T. K. S.: Indian J. Technol. *14,* 151 (1976)
419. Gava, S. A. et al.: Zh. Anal. Khim. *31,* 2129 (1976)
420. Miller, M. P. et al.: Anal. Chem. *49,* 1474 (1977)
421. Gustafson, F. J., Wright J. C.: ibid. *49,* 1680 (1977)
422. Wright, J. C.: ibid. *49,* 1690 (1977)
423. Wright, J. C., Gustafson, F. J.: ibid. *50,* 1147 A (1978)
424. Gustafson, F. J., Wright, J. C.: ibid. *51,* 1762 (1979)
425. Johnston, M. V., Wright, J. C.: ibid. *51,* 1774 (1979)
426. Sevchenko, A. N., Kuznetsova, V. V. in: Redkozemel. Elementi, Moskva, Izd. AN SSSR, 1963, p. 358
427. Melentjeva, E. V., Kononenko, L. I. in: Tez. 2 Vses. Soveshch. po Primen. Org. Reag. v Anal. Khim., Saratov, 1966, p. 147
428. Melentjeva, E. V., Poluektov, N. S., Kononenko, L. I.: Zh. Anal. Khim. *22,* 187 (1967)
429. Melentjeva, E. V., Kononenko, L. I., Poluektov, N. S. in: Sovremen. Metodi Khim. i Spectr. Analiza Mater., Sbornik Obzorov i Metodik, Moskva, Metallurgiya, 1967, p. 202
430. Matveets, M. A., Akhmetova, S. D., Shcherbov, D. P. in: Issled. v Obl. Khim. i Fiz. Metodov Analiza Mineral. Syrya, N 4, Alma-Ata, 1975, p. 32
431. Kononenko, L. I., Poluektov, N. S., Nikonova, M. P.: Zavodsk. Lab. *30,* 779 (1964)
432. Galkina, L. L., Streltsova, S. A.: Zh. Anal. Khim. *26,* 1764 (1971)
433. Williams, D. E., Guyon, J. C.: Mikrochim. Acta N 2, 194 (1972)
434. Fisher, R. P., Winefordner, J. D.: Anal. Chem. *43,* 454 (1971)
435. Shigematsu, T., Matsui, M., Wake, R.: Anal. Chim. Acta *46,* 101 (1969)
436. Karaseva, E. T., Karasev, V. E.: Koord. Khim. *1,* 926 (1975)
437. Karaseva, E. T. et al.: ibid. *5,* 647 (1979)
438. Metlay, M.: J. Electrochem. Soc. *111,* 1253 (1964)
439. Khomenko, V. S., Kuznetsova, V. V., Pekarskaya, L. A.: Zh. Prikl. Spectr. *6,* 117 (1967)
440. Belcher, R., Perry, R., Stephen, W. I.: Analyst *94,* 26 (1969)
441. Tishchenko, M. A. et al.: Zavodsk. Lab. *39,* 671 (1973)
442. Poluektov, N. S., Vitkun, R. A., Kononenko, L. I.: Ukr. Khim. Zh. *30,* 629 (1964)
443. Poluektov, N. S., Kononenko, L. I., Vitkun, R. A. in: Poluchenie i Analiz Veshch. Osoboi Chist., Moskva, Nauka, 1966, p.217
444. Fabian-Mohai, M., Upor, E., Nagy, G.: Acta Chim. Acad. Sci. Hung. *68,* 1 (1971)
445. Kirillov, A. I., Vlasov, N. A., Igumnova, L. I. in: Tez. 8 Konf. Zavodsk. i Promysh. Lab. Kazakhstana i Srednej Azii, Alma-Ata, 1968, p. 72
446. Kononenko, L. I., Lauer, R. S., Poluektov, N. S.: Zh. Anal. Khim. *18,* 1468 (1963)

447. Nikolaeva, K. I., Bozhevolnov, E. A. in: Metodi Analiza Khim. Reakt. i Preparatov., Lumin. Metodi Opred. Microkol. Elementov, N 11, Moskva, IREA, 1965, p. 32
448. Melentjeva, E. V. et al.: Khimichna Promislovist' *3*, 63 (1965)
449. Kononenko, L. I., Melentjeva, E. V., Poluektov, N. S. in: Khim. Transuran. i Oskolochn. Elementov, Leningrad, Nauka, 1967, p. 156
450. Williams, D. E., Guyon, J. C.: Anal. Chem. *43*, 139 (1971)
451. Kachin, S. V., Golovina, A. P., Runov, V. K. in: Org. Reag. v Anal. Khim., Tez. Dokl. Respubl. Konf., Baku, Azerb. Ped. Inst. im V. I. Lenina, 1979, p. 33
452. Kallistratos, G., Pfau, A., Ossowski, B.: Naturwissenschaften *47*, 468 (1960)
453. Tishchenko, M. A. et al.: Zh. Anal. Khim. *28*, 1954 (1973)
454. Tishchenko, M. A. et al. in: Tr. po Khim. i Khim. Tehnol., Fiz.-khim. Metodi Issled. i Analiza, N 3 (34), Gor'kij, 1973, p. 112
455. Shigematsu, T., Matsui, M., Sumida, T.: Bull. Inst. Chem. Res. Kyoto Univ. *46*, 249 (1968)
456. Mishchenko, V. T., Tselik, E. I., Koev, A. P.: Zh. Anal. Khim. *32*, 71 (1977)
457. Karaseva, E. T. et al. in: Stroenie, Svojstva i Primenenie β-Diketonatov, Moskva, Nauka, 1978, p. 165.
458. Kononenko, L. I., Mishchenko, S. A., Poluektov, N. S.: Zh. Anal. Khim. *21*, 1392 (1966)
459. Tishchenko, M. A., Alakaeva, L. A., Poluektov, N. S.: Ukr. Khim. Zh. *37*, 591 (1971)
460. Poluektov, N. S., Tishchenko, M. A., Gerasimenko, G. I.: Zh. Anal. Khim. *30*, 1325 (1975)
461. Tishchenko, M. A., Zheltvaj, I. I., Poluektov, N. S.: Zh. Neorg. Khim. *18*, 2390 (1973)
462. Butter, E., Kolowos, I., Holzapfel, H.: Talanta *15*, 901 (1968)
463. Poluektov, N. S., Alakaeva, L. A., Tishchenko, M. A.: Zavodsk. Lab. *37*, 1077 (1971)
464. Taketatsu, T., Yoshida, S.: Bull. Chem. Soc. Jap. *45*, 2921 (1972)
465. Tishchenko, M. A. et al.: Zh. Anal. Khim. *33*, 2368 (1978)
466. Poluektov, N. S., Alakaeva, L. A., Tishchenko, M. A.: Ukr. Khim. Zh. *38*, 175 (1972)
467. Poluektov, N. S., Alakaeva, L. A., Tishchenko, M. A.: Zh. Anal. Khim *28*, 1621 (1973)
468. Kononenko, L. I., Tishchenko, M. A., Poluektov, N. S.: ibid. *19*, 829 (1964)
469. Dagnall, R. M., Smith, R., West, T. S.: Analyst *92*, 358 (1967)
470. Poluektov, N. S., Alakaeva, L. A., Tishchenko, M. A.: Zh. Anal. Khim. *25*, 2351 (1970)
471. Tishchenko, M. A., Alakaeva, L. A., Poluektov, N. S.: Ukr. Khim. Zh. *39*, 482 (1973)
472. Poluektov, N. S., Tishchenko, M. A., Alakaeva, L. A. in: Tr. po Khim. i Khim. Tekhnol., Poluchenie i Analiz Chist. Veshch., N 4 (35), Gor'kij, 1973, p. 104
473. Tishchenko, M. A., Gerasimenko, G. I., Poluektov, N. S.: Zavodsk. Lab. *40*, 935 (1974)
474. Tishchenko, M. A., Poluektov, N. S.: Ukr. Khim. Zh. *32*, 733 (1966)
475. Kirillov, A. I., Rybakova, M. M.: Zavodsk. Lab. *43*, 1432 (1977)
476. Tishchenko, M. A. et al.: Ukr. Khim. Zh. *32*, 508 (1966)
477. Van Uitert, L. G.: J. Electrchem. Soc. *107*, 803 (1960)
478. Rao, V. J., Rao, D. R., Sinha, A. P. B.: Indian. J. Chem. *8*, 270 (1970)
479. Sato, S., Wada, M.: Bull. Chem. Soc. Jap. *43*, 1955 (1970)
480. Poluektov, N. S. et al.: Opt. i Spectr. *17*, 73 (1964)
481. Poluektov, N. S. et al.: Zh. Prikl. Spectr. *5*, 625 (1966)
482. Kleinerman, M., Choi, S.: J. Chem. Phys. *49*, 3901 (1968)
483. Antipenko, B. M., Privalova, T. A.: Koord. Khim. *4*, 1149 (1978)
484. Belyj, M. U., Kushnirenko, I. Ya.: Zh. Prikl. Spectr. *9*, 272 (1968)
485. Solovjev, E. A., et al.: Vestn. Mosk. Univ. Ser. Khim., N 5, 89 (1966)
486. Bozhevolnov, E. A., Solovjev, E. A., Fakeeva, O. A.: Intern. Conf. Lumin., D) Special Probl. Lumin., 11. Appl. of Lumin., Prepr., Budapest, 1966, p. 41
487. Bozhevolnov, E. A., Solovjev, E. A. in: Tr. IREA, Khim. Reakt. i Preparati, N 30, Moskva, IREA, 1967, p. 202
488. Kirkbright, G. F., Saw, C. G., West, T. S.: Talanta *16*, 65 (1969)
489. Belyj, M. U., Kushnirenko, I. Ya.: Zh. Prikl. Spectr. *10*, 810 (1969)
490. Solovjev, E. A., Bozhevolnov, E. A. in: Tr. IREA, Khim. Reakt. i Preparati, N 26, Moskva, IREA, 1964, p. 194
491. Savvina, L. P., Golovina, A. P., Solovjev, E. A.: Vestn. Mosk. Univ. Ser. Khim. N 4, 451 (1975)
492. Solovjev, E. A. et al.: Zh. Anal. Khim. *30*, 103 (1975)
493. Bozhevolnov, E. A., Solovjev, E. A., Lebedeva, N. A.: ibid. *26*, 1117 (1971)

494. Golovina, A. P., Runov, V. K., Savvina, L. P.: Vestn. Mosk. Univ. Ser. Khim. *18*, 370 (1977)
495. Bozhevolnov, E. A., Solovjev, E. A.: Zh. Anal. Khim. *20*, 1330 (1965)
496. Belyj, M. U., Kushnirenko, I. Ya.: Zh. Prikl. Spectr. *10*, 84 (1969)
497. Belyj, M. U., Kushnirenko, I. Ya.: ibid. *9*, 442 (1968)
498. Kirkbright, G. F. et al.: Talanta *16*, 1081 (1969)
499. Savvina, L. P. et al.: Zh. Anal. Khim. *31*, 1268 (1976)
500. Kirkbright, G. F., Saw, C. G., West, T. S.: Analyst. *94*, 457 (1969)
501. Kondilenko, I., Shishlovsky, A. A.: Dokl. AN SSSR *35*, 264 (1942)
502. Belyj, M. U., Kondilenko, I. I., Shishlovsky, A. A. in: Pamyati S. I. Vavilova, Moskva, Izd. AN SSSR, 1952, p. 247
503. Belyj, M. U., Shishlovsky, A. A.: Izv. An SSSR. Ser. Fiz. *20*, 574 (1956)
504. Belyj, M. U., Rudko, B. F.: Ukr. Fiz. Zh. *5*, 800 (1960)
505. Belyj, M. U., Rudko, B. F.: Izv. AN SSSR. Ser. Fiz. *24*, 582 (1960)
506. Belyj, M. U., Okhrimenko, B. A.: Ukr. Fiz. Zh. *9*, 1059 (1964)
507. Belyj, M. U., Okhrimenko, B. A.: ibid. *9*, 1086 (1964)
508. Belyj, M. U., Kushnirenko, I. Ya.: Izv. AN SSSR. Ser. Fiz. *29*, 387 (1965)
509. Avramenko, V. G., Belyj, M. U.: ibid. *32*, 1401 (1968)
510. Kirkbright, G. F., Saw, C. G., West, T. S.: Analyst *94*, 538 (1969)
511. Belyj, M. U., Kushnirenko, I. Ya., Sheremet, G. P.: Opt. i Spectr. *27*, 90 (1969)
512. Belyj, M. U., Okhrimenko, B. A., Subbota-Melnik, P. A. in: Fiz. Shchelochnogalloid. Kristall., Tr. 2 Vses. Soveshch., Riga, 1962, p. 123
513. Belyj, M. U., Kushnirenko, I. Ya., Okhrimenko, B. A. in: Tr. Komiss. po Spectr. AN SSSR, Mater. 15 Soveshch. po Spectr., Vol. 1, N 1, Moskva, 1964, p. 597
514. Avramenko, V. G., Belyj, M. U., Krivenko, P. I. in: Spectr. Atom. i Mol., Kiev, Naukova Dumka, 1969, p. 313
515. Belyj, M. U., Okhrimenko, B. A.: Opt. i Spectr. *26*, 977 (1969)
516. Gordon, B. E., Shishlovsky, A. A.: Fiz. Zap. AN UkrSSR *8*, 91 (1939)
517. Pringsheim, P., Vogels, H.: Physica *7*, 225 (1940)
518. Gordon, B. E.: Zh. Fiz. Khim. *15*, 448 (1941)
519. Belyj, M. U., Okhrimenko, B. A.: Izv. AN SSSR. Ser. Fiz. *27*, 666 (1963)
520. Belyj, M. U., Kushnirenko, I. Ya.: ibid. *27*, 661 (1963)
521. Belyj, M. U., Okhrimenko, B. A.: ibid. *29*, 391 (1965)
522. Belyj, M. U., Kushnirenko, I. Ya.: Ukr. Fiz. Zh. *9*, 1306 (1964)
523. Belyj, M. U., Kushnirenko, I. Ya. in: Fiz. Shchelochnogalloid. Kristall, Tr. 2 Vses. Soveshch., Riga, 1962, p. 164
524. Belyj, M. U., Kushnirenko, I. Ya.: Ukr. Fiz. Zh. *9*, 1248 (1964)
525. Belyj, M. U., Kushnirenko, I. Ya., Leontev, A. B.: Opt. i Spectr. *34*, 715 (1973)
526. Williams, F. E.: J. Chem. Phys. *19*, 457 (1951)
527. Johnson, P.D., Williams, F. E.: ibid. *20*, 124 (1952)
528. Pekar, S. I.: Uspekhi Fiz. Nauk. *50*, 197 (1953)
529. Yuster, P. H., Delbecq, C. J.: J. Chem. Phys. *21*, 892 (1953)
530. Klement, F. D. in: Tr. Inst. Fiz. i Astron. AN EsSSR, Issled. po Lumin., N 1, Tartu, 1955, p. 3
531. Lushchik, Ch. B. in: Tr. Inst. Fiz. i Astron. AN EsSSR, Issled. Tsentrov Zakhvata v Shchelochnogalloid. Cristallofosfor., N 3, Tartu, 1955
532. Kirs, Ya. Ya. in: Tr. Inst. Fiz. i Astron. AN EsSSR, Issled. po Lumin., N 4, Tartu, 1956, p. 108
533. Lushchik, N. E., Lushchik, Ch. B. in: Tr. Inst. Fiz. i Astron. AN EsSSR, Issled. po Lumin., N 6, Tartu, 1957, p. 5
534. Lushchik, N. E. in: ibid., p. 149
535. Lushchik, N. E. in: Tr. Inst. Fiz. i Astron. AN EsSSR, Issled. po Lumin., N 7, Tartu, 1958, p. 119
536. Shvarts, K. K. in: ibid., p. 153
537. Lushchik, Ch. B., Lushchik, N. E., Shvarts, K. K. in: Tr. Inst. Fiz. i Astron. AN EsSSR, Issled. po Lumin., N 8, Tartu, 1958, p. 3
538. Lushchik, N. E. in: Tr. Inst. Fiz. i Astron. AN EsSSR, Issled. po Lumin., N 10, Tartu, 1959, p. 68
539. Lushchik, N. E., Lushchik, Ch. B.: Opt. i Spectr. *8*, 839 (1960)

540. Lushchik, Ch. B., Lushchik, N. E., Shvarts, K. K.: ibid. *9*, 215 (1960)
541. Zazubovich, S. G., Lushchik, N. E. in: Tr. Inst. Fiz. i Astron. AN EsSSR, Issled. po Lumin., N 14, Tartu, 1961, p. 283
542. Lushchik, Ch. B. et al. in: Fiz. Shchelochnogalloid. Kristall., Tr. 2 Vses. Soveshch., Riga, 1962, p. 102
543. Golovina, A. P. et al. in: Khim. i Tekhnol. Khalkogenov i Khalkogenidov, Tez. 1 Vses. Soveshch., Karaganda, 1978, p. 284
544. De Armond, K., Forster, L. S.: Spectrochim. Acta *19*, 1393 (1963)
545. De Armond, K., Forster, L. S.: ibid. *19*, 1403 (1963)
546. De Armond, K., Forster, L. S.: ibid. *19*, 1687 (1963)
547. Schläfer, H. L., Gausmann, H., Witzke, H.: Z. Phys. Chem. (N. F.) *56*, 55 (1967)
548. Pfeil, A.: J. Amer. Chem. Soc. *93*, 5395 (1971)
549. Kupka, J., Lukowiak, E., Jezowska-Trzebiatowska, B.: Spectrochim. Acta *28A*, 1987 (1972)
550. Schmidtke, H. H., Hoggard, P. E.: Chem. Phys. Lett. *20*, 119 (1973)
551. Mc Caffery, A. J. et al.: ibid. *22*, 600 (1973)
552. Castelli, F., Forster, L. S.: J. Amer. Chem. Soc. *95*, 7223 (1973)
553. Kataoka, H.: Bull. Chem. Soc. Jap. *46*, 2078 (1973)
554. Hoggard, P. E., Schmidtke, H. H.: Inorg. Chem. *12*, 1986 (1973)
555. Kutal, C., Adamson, A. W.: ibid. *12*, 1990 (1973)
556. De Armond, M. K.: Acc. Chem. Res. *7*, 309 (1974)
557. Sabbatini, N., Scandola, M. A., Balzani, V.: J. Phys. Chem. *78*, 541 (1974)
558. Kane-Maguire, N. A. P., Conway, J., Langford, C. H.: J. Chem. Soc. Chem. Comm. *19*, 801 (1974)
559. Otto, H., Yersin, H., Gliemann, G.: Z. Phys. Chem. (N. F.) *92*, 193 (1974)
560. Weber, M. J., Varitimos, T. E.: J. Appl. Phys. *45*, 810 (1974)
561. Castelli, F., Forster, L. S.: Phys. Rev. *11B*, 920 (1975)
562. Watson, W. M. et al.: Inorg. Chem. *14*, 2374 (1975)
563. Solovjev, E. A., Tikhonov, G. P., Bozhevolnov, E. A.: Zh. Prikl. Spectr. *23*, 434 (1975)
564. Solovjev, E. A. et al.: Zavodsk. Lab. *39*, 669 (1973)
565. Solovjev, E. A. et al. in: Metodi Khim. Anliza, Moskva, Mosk. Dom Nauch.-tekhn. Propagandi im F. E. Dzerzhinskogo, 1969, p. 184
566. Khacheryan, S. L., Solovjev, E. A. in: Reakt. i Osobo Chist. Veshch., N 5, Moskva, Nauch.-issled. Inst. Tekhn.-ekon. Issled., 1978, p. 24
567. Mc Glynn, S. P., Smith, J. K.: J. Mol. Spectr. *6*, 164 (1961)
568. Mc Glynn, S. P., Smith, J. K., Neely, W. C.: J. Chem. Phys. *35*, 105 (1961)
569. Volodko, L. V., Komyak, A. I., Sleptsov, L. E.: Opt. i Spectr. *23*, 730 (1967)
570. Volodko, L. V., Sevchenko, A. N., Umrejko, D. S.: Dokl. AN SSSR *135*, 560 (1960)
571. Volodko, L. V., Sevchenko, A. N., Umrejko, D. S.: Izv. AN SSSR. Ser. Fiz. *24*, 749 (1960)
572. Bartecki, A., Jezowska-Trzebiatowska, B.: Nucleonika *6*, 267, 277, 287 (1961)
573. Belford, R. L., Belford, G.: J. Chem. Phys. *34*, 1330 (1961)
574. Jezowska-Trzebiatowska, B., Bartecki, A.: Spectrochim. Acta *18*, 799 (1962)
575. Volodko, L. V., Sevchenko, A. N., Umrejko, D. S.: Izv. AN SSSR. Ser. Fiz. *27*, 651 (1963)
576. Volodko, L. V., Sevchenko, A. N., Umrejko, D. S.: Opt. i Spectr. *17*, 356 (1964)
577. Bell, J. T., Biggers, R. E.: J. Mol. Spectr. *18*, 247 (1965)
578. Newman, J. B.: J. Chem. Phys. *43*, 1691 (1965)
579. De Jaegere, S., Govers, T.: Nature *205*, 900 (1965)
580. Baran, V.: Coll. Czech. Chem. Comm. *31*, 2093 (1966)
581. Volodko, L. V., Sevchenko, A. N., Umrejko, D. S.: Dokl. AN SSSR *172*, 1303 (1967)
582. Bell, J. T., Biggers, R. E.: J. Mol. Spectr. *25*, 312 (1968)
583. Umrejko, D. S., Larkin, G. N.: Zh. Prikl. Spectr. *8*, 447 (1968)
584. De Jaegere, S., Gorller-Walrand, C.: Spectrochim. Acta *25A*, 559 (1969)
585. Novitsky, G. G., Sevchenko, A. N., Umrejko, D. S.: Dokl. AN SSSR *187*, 1268 (1969)
586. Gorller-Walrand, C., Vanquickenborne, L. G.: J. Chem. Phys. *54*, 4178 (1971)
587. Sergeeva, G. I. et al.: Zh. Prikl. Spectr. *19*, 419 (1973)
588. Sergeeva, G. I. et al.: Khim. Vys. Energ. *8*, 38 (1974)
589. Dobrolyubskaya, T. S., Anikina, L. I.: Zh. Anal. Khim. *22*, 1841 (1967)
590. Dobrolyubskaya, T. S.: ibid. *26*, 926 (1971)

591. Pakalns, P.: Austr. At. Energy Comm., AAEC/TM (Rep.), AAEC/TM-552, *6*, 1, 5 (1970)
592. Huffman, C., Riley, L. B.: U. S. Geol. Surv., Prof. Pap. N 700-B *181* (1970)
593. Basyrova, A. N., Nikiforov, V. S. in: Radioekol. Issled. v Prirod. Biogeotsen., Moskva, Nauka, 1972, p. 238
594. Gopinathan, C., Stevens, G., Hart, E. J.: J. Phys. Chem. *76*, 3698 (1972)
595. Danielsson, A. et al.: Talanta *20*, 185 (1973)
596. Zolin, V. F., Rozman, S. P., Fisher, P. S.: Opt. i Spectr. *35*, 589 (1973)
597. Szoke, J., Szilagyi, I.: Anal. Chem. 46, 292 (1974)
598. Pant, D. D., Tripathi, H. B.: J. Lumin. *8*, 492 (1974)
599. Sergeeva, G. et al.: J. Chem. Soc. Chem. Comm., N5, 159 (1974)
600. Blasse, G.: Chem. Phys. Lett. *44*, 61 (1976)
601. Pakalns, P., Ismay, L. E.: Mikrochim. Acta *1*, N2–3, 297 (1976)
602. Pakalns, P., Ismay, L. E.: ibid. *2*, N1–2, 217 (1976)
603. Bykhovtsova, T. T.: Zavodsk. Lab. *43*, 1041 (1977)
604. Konstantinova, M. H., Mareva, S. M., Yordanov, N.: Dokl. Bolg. AN *31*, 205 (1978)
605. Shlyapintokh, V. Ya. et al.: Chemiluminescence Methods of Investigation Slow Chemical Processes, Moscow, Nauka, 1966
606. Seitz, W. R., Hercules, D. M.: Int. J. Environ. Anal. Chem. *2*, 273 (1973)
607. Isacsson, U., Wettermark, G.: Anal. Chim. Acta *68*, 339 (1974)
608. Seitz, W. R., Neary, M. P.: Anal. Chem. *46*, 188A (1974)
609. Hastings, J. W., Wilson, T.: Photochem. Photobiol. *23*, 461 (1976)
610. Paul, D. B.: Talanta *25*, 377 (1978)
611. Babko, A. K., Dubovenko, L. I., Lukovskaya, N. M.: Chemiluminescence Analysis, Kiev, Tekhnika, 1966
612. Cormier, M. J., Hercules, D. M., Lee, J. (eds.): Chemiluminescence and Bioluminescence, New York, Plenum, 1973
613. Pantel, S., Weisz, H.: Anal. Chim. Acta *74*, 275 (1975)
614. Bowling, J. L. et al.: ibid. *76*, 47 (1975)
615. Hoyt, S. D., Ingle, J. D.: ibid. *87*, 163 (1976)
616. Tiffany, T. O. in: Modern Fluorescence Spectroscopy, New York, Plenum, 1976, p. 1
617. Steig, S., Nieman, T. A.: Anal. Chem. *49*, 1322 (1977)
618. Nau, V., Nieman, T. A.: ibid. *51*, 424 (1979)
619. Rigin, V. I., Blokhin, A. I.: Zh. Anal. Khim. *32*, 312 (1977)
620. Seitz, W. R., Suydam, W. W., Hercules, D. M.: Anal. Chem. *44*, 957 (1972)
621. Li, R. T., Hercules, D. M.: ibid. *46*, 916 (1974)
622. Rigin, V. I., Bakhmurov, A. S.: Zh. Anal. Khim. *31*, 93 (1976)
623. Lukovskaya, N. M., Bilochenko, V. A.: ibid. *32*, 2177 (1977)
624. Lukovskaya, N. M., Bilochenko, V. A.: Ukr. Khim. Zh. *43*, 756 (1977)
625. Lukovskaya, N. M., Terletskaya, A. V., Bogoslovskaya, T. A.: Zh. Anal. Khim. *29*, 2268 (1974)
626. Lukovskaya, N. M., Bogoslovskaya, T. A.: Ukr. Khim. Zh. *41*, 200 (1975)
627. Pilipenko, A. T., Angelova, G. V., Kalinichenko, I. E.: ibid. *40*, 1302 (1974)
628. Lukovskaya, N. M., Bogoslovskaya, T. A.: ibid. *41*, 268 (1975)
629. Lukovskaya, N. M., Bogoslovskaya, T. A.: ibid. *41*, 529 (1975)
630. Dubovenko, L. I., Korotun, L. M.: Vestn. Kiev. Univ. Ser. Khim. N15, 12 (1974)
631. Dubovenko, L. I., Tananajko, M. M., Drokov, V. G.: Ukr. Khim. Zh. *40*, 758 (1974)
632. Dubovenko, L. I., Beloshitsky, N. V.: Zh. Anal. Khim. *29*, 111 (1974)
633. Montano, L. A., Ingle, J. D.: Anal. Chem. *51*, 919 (1979)
634. Montano, L. A., Ingle, J. D.: ibid. *51*, 926 (1979)
635. Terletskaya, A. V., Lukovskaya, N. M., Anatienko, N. L.: Zh. Anal. Khim. *34*, 1460 (1979)
636. Dubovenko, L. I., Guta, A. M.: Izv. Vyssh. Uchebn. Zaved. Ser. Khim. i Khim. Tekhnol. *18*, 1211 (1975)
637. Guta, A. M.: Vestn. L'vov. Univ. Ser. Khim., N15, 40 (1974)
638. Drokov, V. G., Dubovenko, L. I.: Ukr. Khim. Zh. *40*, 549 (1974)
639. Tovmasyan, A. P., Dubovenko, L. I.: Molod. Nauch. Rabotn. Ser. Estestv. Nauk, N2(18), 100 (1973)
640. Truba, N. A., Nabivanets, B. I.: Gidrobiol. Zh. *11*, N2, 125 (1975)

641. Dubovenko, L. I., Nazarenko, A. Yu.: Ukr. Khim. Zh. *41*, 1205 (1975)
642. Dubovenko, L. I., Bilochenko, V. A.: ibid. *40*, 423 (1974)
643. Lukovskaya, N. M., Terletskaya, A. V., Kushchevskaya, N. F.: Zh. Anal. Khim. *33*, 750 (1978)
644. Lukovskaya, N. M., Markova, L. V., Evtushenko, N. F.: ibid. *29*, 767 (1974)
645. Lukovskaya, N. M., Kushchevskaya, N. F.: Ukr. Khim. Zh. *42*, 87 (1976)
646. Nabivanets, B. I., Truba, N. A.: Gidrobiol. Zh. *9*, N 5, 90 (1973)
647. Hasegawa, A., Somiya, T., Niki, E.: Kogakuin Daigaku Kenkyu Hokoku *37*, 119 (1974)
648. Dubovenko, L. I., Nazarenko, A. Yu.: Zh. Anal. Khim. *32*, 1345 (1977)
649. Dubovenko, L. I., Evtushenko, N. F.: Vestn. Kiev. Univ. Ser. Khim., N 14, 12 (1973)
650. Pilipenko, A. T. et al.: Ukr. Khim. Zh. *40*, 1205 (1974)
651. Skorobogatyj, Ya. P., Zinchuk, V. K.: Zh. Anal. Khim. *30*, 819 (1975)
652. Zinchuk, V. K., Skorobogatyj, Ya. P. in: Tr. po Khim. i Khim. Tekhnol., Poluchenie i Analiz Chist. Veshch., N 4 (35), Gor'kij, 1973, p. 117
653. Lukovskaya, N. M., Teletskaya, A. V.: Ukr. Khim. Zh. *40*, 1311 (1974)
654. Lukovskaya, N. M., Teletskaya, A. V.: ibid. *40*, 1193 (1974)
655. Lukovskaya, N. M., Bilochenko, V. A.: Zavodsk. Lab. *40*, 936 (1974)
656. Rigin, V. I., Bakhmurov, A. S., Blokhin, A. I.: Zh. Anal. Khim. *30*, 2413 (1975)
657. Lukovskaya, N. M., Kushchevskaya, N. F.: Ukr. Khim. Zh. *41*, 643 (1975)
658. Koshcheeva, I. Ya., Varshal, G. M., Ejger, V. I. in: Novye Metodi Vydel. i Opred. Blagorodn. Elementov, Moskva, GEOKhI im. V. I. Vernadskogo AN SSSR, 1975, p. 89
659. Lukovskaya, N. M., Terletskaya, A. V.: Zh. Anal. Khim. *31*, 751 (1976)
660. Zinchuk, V. K., Rekhlitskaya, L. M.: Vestn. L'vov. Univ. Ser. Khim. N 15, 35 (1974)
661. Pilipenko, A. T., Mitropolitska, E. V., Lukovskaya, N. M.: Ukr. Khim. Zh. *41*, 525 (1975)
662. Pilipenko, A. T., Mitropolitska, E. V., Lukovskaya, N. M.: ibid. *41*, 1196 (1975)
663. Pilipenko, A. T., Mitropolitska, E. V., Lukovskaya, N. M.: ibid. *39*, 73 (1973)
664. Schenk, G. H., Dilloway, K. P.: Anal. Lett.*2*, 379 (1969)
665. Taves, D. R.: Talanta *15*, 1015 (1968)
666. Tan Lay, Har, West, T. S.: Anal. Chem. *43*, 136 (1971)
667. Guyon, J. C., Jones, B. E., Britton, D. A.: Mikrochim. Acta, N 6, 1180 (1968)
668. McKinney, G. L., Lau, H. K. Y., Lott, P. F.: Microchem. J. *17*, 375 (1972)
669. Morgen, E. A., Vlasov, N. A., Tyutin, V. A. in: Gidrokhim. Mater., Khim. Prirod. Vod, ikh Zagryazn. i Samoochishch., Vol. 50, Leningrad, Gidrometeorolog. Izd., 1969, p 92
670. Shafran, I. G. et al. in: Metodi Analiza Khim. Reakt. i Preparatov, N 21, Moskva, IREA, 1973, p. 123
671. Lay Har, Tan, West, T. S.: Analyst *96*, 281 (1971)
672. Vernon, F., Whitham, P.: Anal. Chim. Acta *59*, 155 (1972)
673. Bark, L. S., Rixon, A.: Analyst *95*, 786 (1970)
674. Axelrod, H. D., Engel, N. A.: Anal. Chem. *47*, 922 (1975)
675. Axelrod, H. D., Pickett, R. A., Engel, N. A.: ibid. *47*, 2021 (1975)
676. Dombrowski, L. J., Pratt, E. J.: ibid. *44*, 2268 (1972)
677. Wiersma, J. H.: Anal. Lett. *3*, 123 (1970)
678. Sawicki, C. R.: ibid. *4*, 761 (1971)
679. Afghan, B. K., Ryan, J. F.: Anal. Chem. *47*, 2347 (1975)
680. Nakano, S. et al.: Chem. Pharm. Bull. *25*, 1237 (1977)
681. Kramer, D. N.: Pure Appl. Chem. *48*, 65 (1976)
682. Nakamura, H., Tamura, Z.: Chem. Pharm. Bull. *22*, 1950 (1974)
683. Nakamura, H., Tamura, Z.:: ibid. *23*, 1261 (1975)
684. Axelrod, H. D., Bonelli, J. E., Lodge, J. P.: Anal. Chim. Acta *51*, 21 (1970)
685. Axelrod, H. D., Bonelli, J. E., Lodge, J. P.: Anal. Chem. *42*, 512 (1970)
686. Parker, C. A., Harvey, L. G.: Analyst *87*, 558 (1962)
687. Parker, C. A., Harvey, L. G.: ibid. *86*, 54 (1961)
688. Wheeler, G. L., Lott, P. F.: Microchem. J. *19*, 390 (1974)
689. Crenshaw, G. L., Lakin, H. W.: J. Res. U. S. Geol. Surv. *2*, 483 (1974)
690. Haddad, P. R., Smythe, L. E.: Talanta *21*, 859 (1974)
691. Nazarenko, I. I. et al.: Zh. Anal. Khim. *30*, 733 (1975)
692. Chan, C. C. Y.: Anal. Chim. Acta *82*, 213 (1976)

693. Shawky, M., White, C. L.: Anal. Chem. *48*, 1484 (1976)
694. Brown, M. W., Watkinson, J. H.: Anal. Chim. Acta *89*, 29 (1977)
695. Guilbault, G. G., Brignac, P., Zimmer, M.: Anal. Chem. *40*, 190 (1968)
696. Axelrod, H. D., Bonelli, J. E., Lodge, J. P.: Environ. Sci. Technol. *5*, 420 (1971)
697. Shcherbov, D. P., Lisitsina, D. N. in: Issled. v Obl. Khim. i. Fiz. Metodov Analiza Mineral. Syrya, N4, Alma-Ata, 1975, p. 74
698. Colovos, G., Haro, M., Freiser, H.: Talanta *17*, 273 (1970)
699. Stolyarov, K. P., Grigorjev, N. N., Khomenok, G. A.: Vestn. Leningrad Univ. Ser. Fiz. i Khim., N22, 120 (1972)
700. Cerda, M. V., Mongay, F. C.: Quim. Anal. (pura y apl.) *30*, 15 (1976)
701. Yamamoto, D., Kisu, K.: Bunseki Kagaku *23*, 638 (1974)
702. Kirkbright, G. F., Narayanaswamy, R., West, T. S.: Anal. Chem. *43*, 1434 (1971)
703. Podberezskaya, N. K., Sushkova, V. A.: Zavodsk. Lab. *39*, 774 (1973)
704. Podberezskaya, N. K., Sushkova, V. A., Shilenko, E. A. in: Issled. v Obl. Khim. i Fiz. Metodov Analiza Mineral. Syrya, Alma-Ata, 1971, p. 87
705. Busljeta, M., Weber, K.: Acta Pharm. Jugosl. *25*, 247 (1975)
706. Lukovskaya, N. M., Bogoslovskaya, T. A.: Zh. Anal. Khim. *29*, 674 (1974)
707. Lukovskaya, N. M., Anatienko, N. L. in: Org. Reag. v Anal. Khim., Tez. 4 Vses. Konf.,Vol. 2, Kiev, Naukova Dumka, 1976, p. 80
708. Bishop, E. (ed.): Indicators, Internat. Ser. Monogr. Analyt. Chem., Vol. 51, Oxford, Pergamon, 1972
709. Andrushko, G. S., Maksimycheva, Z. T., Talipov, Sh. T.: Uzb. Khim. Zh. N2, 24 (1969)
710. Andrushko, G. S., Talipov, Sh. T., Maksimycheva, Z. T.: ibid. N5, 32 (1970)
711. Andrushko, G. S., Maksimycheva, Z. T., Talipov, Sh. T.: Dokl. AN UzSSR, N3, 26 (1970)
712. Borle, A. B., Briggs, F. N.: Anal. Chem. *40*, 339 (1968)
713. Escarrilla, A. M.: Talanta *13*, 363 (1966)
714. Bermejo-Martinez, F., de Lopidana, G.: Anal. Chim. Acta *47*, 139 (1969)
715. Huitink, G. M., Diehl, H.: Talanta *21*, 1193 (1974)
716. Beck, J. L., Fitzgerald, J. M., Bishop, J. A.: Anal.Chim. Acta *51*, 191 (1970)
717. Van Slageren, R., den Boef, G., van der Linden, W. E.: Talanta *20*, 739 (1973)
718. Stolyarov, K. P., Firyulina, V. V., Gabajdulin, I. Z.: Zavodsk. Lab. *38*, 664 (1972)
719. Solovjeva, L. A., Stolyarov, K. P., Grigorjev, N. N. in: Instrument. i Khim. Metodi Analiza, Leningrad, Izd. Leningrad. Univ., 1973, p. 98
720. Van der Linden, W. E., den Boef, G., Ozinga, W.: Mikrochim. Acta *1*, N1, 83 (1976)
721. Vladimirova, V. M. et al.: Zavodsk. Lab. *32*, 1045 (1966)
722. Singhal, G. K., Tandon, K. N.: Talanta *14*, 1351 (1967)
723. Lukovskaya, N. M., Markova, L. V.: Zh. Anal. Khim. *24*, 1893 (1969)
724. Babko, A. K., Lukovskaya, N. M.: Ukr. Khim. Zh. *35*, 1060 (1969)
725. Erdey, L., Weber, O., Buzas, I.: Talanta *17*, 1221 (1970)
726. Sarudi, I.: Z. Anal. Chem. *260*, 114 (1972)
727. Grigorenko, F. F., Dubovenko, L. I., Kovalenko, G. I.: Zavodsk. Lab. *39*, 133 (1973)
728. Stuzka, V., Golovina, A. P., Alimarin, I. P.: Coll. Czech. Chem. Comm. *34*, 221 (1969)
729. Temkina, V. Ya., Yaroshenko, G. F., Lastovsky, R. P.: Zh. Anal. Khim. *22*, 632 (1967)
730. Temkina, V. Ya. et al. in: Tr. IREA, Khim. Reakt. i Preparati, N30, Moskva, IREA, 1967, p. 131
731. Maksimycheva, Z. T. et al. in: Tr. Tashkent. Univ., Anal. Khim. Redk. i Rasseyan. Elementov, N323, Tashkent, 1968, p. 96
732. Schulman, S. G., Sturgeon, R. J.: Anal. Chim. Acta *93*, 239 (1977)
733. Schulman, S. G., Underberg, W. J. M.: ibid. *107*, 411 (1979)
734. Roman, M., Ales-Barrero, F.: Quim. Anal. (pyra y apl.) *29*, 301, 316 (1975)
735. Stolyarov, K. P., Firyulina, V. V. in: Opt. Metodi Kontr. Khim. Sost. Mater., Moskva, Mosk. Dom Nauch.-tekhn. Propagandi im. F. E. Dzerzhinskogo, 1974, p. 63
736. Maksimycheva, Z. T. et al. in: Tr. Tashkent. Univ., Anal. Khim. Redk. i Rasseyan. Elementov, N323, Tashkent, 1968, p. 102
737. Talipov, Sh. T. et al. in: ibid. p. 107
738. Salam Khan, M. A., Mooney, E. F., Stephen, W. I.: Anal. Chim. Acta *43*, 153 (1968)
739. Ditz, J., Suk, V., Neumann, J.: Chem. Listy *62*, 1330 (1968)

740. Neumann, J., Ditz, J., Suk, V.: Z. Anal. Chem. *239*, 167 (1968)
741. Temkina, V. Ya. et al.: Zh. Anal. Khim. *24*, 240 (1969)
742. Temkina, V. Ya. et al. in: Tr. IREA, Khim. React. i Preparati, N 32, Moskva, IREA, 1970, p. 60
743. Temkina, V. Ya. et al.: Dokl. AN SSSR *194*, 602 (1970)
744. Clements, R. L., Read, J. I., Sergeant, G. A.: Analyst *96*, 656 (1971)
745. Roman, M., Ales-Barrero, F.: Quim. Anal. (pura y apl.) *29*, 323 (1975)
746. Robinson, J. L., Lott, P. F.: Microchem. J. *19*, 115 (1974)

Author-Index Volumes 1–47

A. F. Williams

A Theoretical Approach to Inorganic Chemistry

1979. 144 figures, 17 tables. XII, 316 pages
ISBN 3-540-09073-8

Contents: Quantum Mechanics and Atomic Theory. – Simple Molecular Orbital Theory. – Structural Applications of Molecular Orbital Theory. – Electronic Spectra and Magnetic Properties of Inorganic Compounds. – Alternative Methods and Concepts. – Mechanism and Reactivity. – Descriptive Chemistry. – Physical and Spectroscopic Methods. – Appendices. – Subject Index.

This book is intended to outline the application of simple quantum mechaniscs to the study of inorganic chemistry, and to show its potential for systematizing and understanding the structure, physical properties, and reactivities of inorganic compounds. The considerable development of inorganic chemistry in recent years necessitates the establishment of a theoretical framework if the student is to acquire sound knowledge of the subject. An effort has been made to cover a wide range of subjects, and to encourage the reader to think of further extensions of the theories discussed. The importance of the critical application of theory is emphasized, and, although the book is concerned chiefly with molecular orbital theory, other approaches are discussed. The book is intended for students in the latter half of their undergraduate studies.

Springer-Verlag
Berlin
Heidelberg
New York

Inorganic Chemistry Concepts

Editors: M. Becke, C. K. Jørgensen, M. F. Lappert, S. J. Lippard, J. L. Margrave, K. Niedenzu, R. W. Parry, H. Yamatera

Volume 1
R. Reisfeld, C. K. Jørgensen
Lasers and Excited States of Rare Earths
1977. 9 figures, 26 tables. VIII, 226 pages
ISBN 3-540-08324-3

Contents: Analogies and Differences Between Monatomic Entities and Condensed Matter. – Rare-Earth Lasers. - Chemical Bonding and Lanthanide Spectra. - Energy Transfer. - Applications and Suggestions.

Volume 2
R. L. Charlin, A. J. van Duyneveldt
Magnetic Propeties of Transition Metal Compounds
1977. 149 figures, 7 tables. XV, 264 pages
ISBN 3-540-08584-X

Contents: Paramagnetism: The Curie Law. - Thermodynamics and Relaxation.- Paramagnetism: Zero-Field Splittings. - Dimers and Clusters. - Lone-Range Order. - Short-Range Order. - Special Topics: Spin-Flop, Metamagnetism, Ferrimagnetism and Canting. - Selected Examples.

Volume 3
P. Gütlich, R. Link, A. Trautwein
Mössbauer Spectroscopy and Transition Metal Chemistry
1978. 160 figures, 1 folding plate, 19 tables. X, 280 pages
ISBN 3-540-08671-4
Contents: Basic Physical Concepts. - Hyperfine Interactions. - Experimental. - Mathematical Evaluation of Mössbauer Spectra. - Interpretation of Mössbauer Parameters of Iron Compounds. - Mössbauer-Active Transition Metals Other than Iron. - Some Special Applications.

Springer-Verlag
Berlin Heidelberg New York

Volume 4
Y. Satio
Inorganic Molecular Dissymmetry
1979. 107 figures, 28 tables. IX, 167 pages
ISBN 3-540-09176-9

Contents: Introduction. - X-Ray Diffraction. - Conformational Analysis. - Structure and Isomerism of Optically Active Complexes.-Electron-Density Distribution in Transition Metal Complexes.- Circular Dichroism. - References. - Subject Index.

Volume 5
T. Tominaga, E. Tachikawa
Modern Hot Atom Chemistry and Its Applications
1981. 57 figures, approx. 34 tables. Approx. 160 pages
ISBN 3-540-10715-0

Contents: Introduction.-Experimental Techniques: Production of Energetic Atoms. - Radiochemical Separation Techniques. - Special Physical Techniques. -Characteristics of Hot Atom Reactions: Gas Phase Hot Atom Reactions. - Applications of Hot Atom Chemistry and Related Topics: Applications in Inorganic, Analytical and Geochemistry. - Applications in Physical Chemistry. - Applications in Biochemistry and Nuclear Medicine. - Hot Atom Chemistry in Energy-related Research. Current Topics Related to Hot Atom Chemistry and Future Scope.

Volume 6
D. L. Kepert
Inorganic Stereochemistry
1981. 206 figures, approx. 45 tables. Approx. 250 pages
ISBN 3-540-10716-9

Contents: Introduction. - Polyhedra. - Four-Coordinate Compounds. - Five Coordinate Compounds Containing Only Unidentate Ligands. - Five-Coordinate Compounds Containing Chelate Groups. - Six-Coordinate Compounds Containing Only Unidentate Ligands. - Six-Coordinate Compounds [M(Bidentate)$_2$ (Unidentate)$_2$]. - Six-Coordinate Compounds [M(Bidentate)$_3$]. - Six-Coordinate Compounds Containing Tridentate Ligands. - Seven-Coordinate Compounds Containing Only Unidentate Ligands. - Seven-Coordinate Compounds Containing Chelate Groups. - Eight-Coordinate Compounds Containing Only Unidentate Ligands. - Eight-Coordinate Compounds Containing Chelate Groups. - Nine-Coordinate Compounds. - Ten-Coordinate Compounds. - Twelve-Coordinate Compounds. -

www.ingramcontent.com/pod-product-compliance
Ingram Content Group UK Ltd.
Pitfield, Milton Keynes, MK11 3LW, UK
UKHW050647051126
16351UKWH00012B/61

9783662153598